COPERNICAN QUESTIONS

A Concise Invitation to the Philosophy of Science

KEITH PARSONS

University of Houston, Clear Lake

Boston   Burr Ridge, IL   Dubuque, IA   Madison, WI   New York
San Francisco   St. Louis   Bangkok   Bogotá   Caracas   Kuala Lumpur
Lisbon   London   Madrid   Mexico City   Milan   Montreal   New Delhi
Santiago   Seoul   Singapore   Sydney   Taipei   Toronto

The **McGraw·Hill** Companies

Mc Graw Hill Higher Education

COPERNICAN QUESTIONS: A CONCISE INVITATION TO THE
PHILOSOPHY OF SCIENCE
Published by McGraw-Hill, an imprint of The McGraw-Hill Companies, Inc.,
1221 Avenue of the Americas, New York, NY 10020. Copyright © 2006. All rights

This book is printed on acid-free paper.

1 2 3 4 5 6 7 8 9 0 FGR/FGR 0 9 8 7 6 5

ISBN 0-07-285020-5

Editor in Chief: Emily Barrosse
Publisher: Lyn Uhl
Sponsoring Editor: Jon-David Hague
Editorial Coordinator: Allison Rona
Signing Representative:
  Paul Moorman
Executive Marketing Manager:
  Suzanna Ellison

Project Manager: Christina Gimlin
Manuscript Editor: Jan Fehler
Design Manager: Cassandra Chu
Interior Designer: Glenda King
Production Supervisor:
  Tandra Jorgensen

Composition: 10/13 New Baskerville by Thompson Type
Printing: 45# New Era Matte, Quebecor Fairfield

Library of Congress Cataloging-in-Publication Data
Parsons, Keith M., 1952–
  Copernican questions : a concise invitation to the philosophy
of science / Keith Parsons.
    p. cm.
  Includes bibliographical references and index.
  ISBN 0-07-285020-5
    1. Science—History—Philosophy. 2. Science—Methodology.
3. Constructive realism. I. Title.
Q174.8.P37 2005
501—dc22   2005052216

www.mhhe.com

CONTENTS ✍

PREFACE FOR
INSTRUCTORS

THIS IS NOT A TYPICAL INTRODUCTION to the philosophy of science. A number of such introductory texts exist and quite a few of them are excellent. This book differs from such more conventional books in aim, content, and tone. My aim is not to introduce the whole field of philosophy of science, and not even to comprehensively survey the topics covered; this is why the book is subtitled an "invitation" rather than an "introduction" to the philosophy of science. I want to invite students to think deeply about some salient issues in the philosophy of science. Typically, a textbook takes up each of the major topics—explanation, confirmation, the nature of theories, and so forth—and then surveys and evaluates the various philosophical opinions on each topic. There is much to be learned from this approach and many of the texts that take it do a very effective job. I have a very different view of the purpose of an introductory text. In my view, an introductory-level work should serve as an appetizer, not a main course. An appetizer serves to whet the appetite, and I find that nothing stimulates beginning students like *controversy*. Therefore, my aim here is not to provide students with an overview, but to confront them with controversy.

I also think that for students taking their first course in philosophy of science it is much more effective to get into greater depth on just a couple of issues than to have a broad survey of the whole field. This book therefore aims to take students rather deeply into just two issues: the rationality of science and the realism question.

For the past four decades, discussions of the nature of scientific rationality have begun with Thomas Kuhn's *The Structure of Scientific Revolutions*.

The present book therefore devotes more space to Kuhn than to any other single philosopher. The entire second chapter deals with Kuhn's various doctrines of incommensurability, which in my view constitute Kuhn's most significant challenge to traditional notions of scientific rationality. In the so-called science wars of the 1990s, much more radical critiques of scientific rationality were furiously debated. Chapter Three examines a few of the most radical critiques of scientific objectivity from the "academic left"—social constructivists, postmodernists, and gender feminists—as well as those arising from the anti-evolutionary right, as represented by Phillip E. Johnson. While the science wars involved academics from many fields and specialties, from physics to comparative literature, the debate over realism was to a much greater degree restricted to professional philosophers of science. Here also a particular book prompted much of the discussion, Bas van Fraassen's *The Scientific Image*. Van Fraassen's book, and other arguments over realism and antirealism, will ground the discussion in Chapters Four and Five. Standard introductory topics such as explanation and confirmation are examined *only* to the extent that they bear on the issues of rationality and realism.

Some of the referees who read the manuscript for this book complained that it left out a number of their favorite topics, such as inductivism vs. falsificationism. Again, my aim is not to offer a survey of the philosophy of science, but to guide students fairly deeply into just a *small* set of core problems. It is crucial to the purpose of this text that its structure be kept simple and focused; adding extra topics would confuse the structure and lose the focus.

The referees who read the manuscript for this book had far more complaints about Chapter Three than any other. One charge was that I take a sarcastic or dismissive tone towards the writers I examine there. I have rewritten the chapter extensively in an effort to remove or reword any passages that might give such an impression. Some readers will no doubt feel that I have failed in that effort and that the text still connotes an attitude of condescension or contempt towards some of the views I consider. For the record, then, let me state that I regard all of the writers examined in this book to be serious scholars who are making serious claims, and who are therefore worthy of the earnest critique I have tried to offer.

Another complaint of several referees was that the particular writers I examine in the third chapter are not the best representatives of social constructivism, postmodernism, or feminism. They charged that I had selected the most extreme and least plausible representatives of those viewpoints. I disagree that the particular authors I chose are unrepresentative of their respective viewpoints. They may state their views more immoderately, with less subtlety, or with fewer qualifications than others, but I do not see that their positions are far outside the "mainstream" writers in those camps. Also, let me emphasize again that the aim of this book is not to offer a survey or overview of any sort. When I mention, for instance, the social constructivists, postmodernists, or feminist philosophers of science, I examine only a very few of these writers and I make no attempt to present a comprehensive survey of their fields. Major figures in these fields, with the exception of one or two individuals, are not mentioned at all. For instance, any survey of feminist philosophy of science would have to deal with Evelyn Fox Keller, Helen Longino, and Kathleen Okruhlik. I mention none of these simply because my aim was not to write a conspectus, but to get into the issues that, in my experience, are most effective in engaging and motivating students. Sandra Harding filled the bill very well here. She makes strong claims and backs them with clever arguments, so I focus on her as the representative of feminist philosophy of science.

Actually, the topics in this book will seem somewhat out of date for professional philosophers. The realism/antirealism issue is presently deadlocked and the issues about Kuhn, incommensurability, and the rationality of science are well-trodden ground. Professional philosophers of science might regard both issues with a bit of ennui. But a textbook, of course, is primarily for the students. A couple of recently published, excellent books offer absolutely up-to-date surveys of current philosophy of science. The only criticism I have of these books is that beginning students would find them utterly incomprehensible. The best way to begin to understand the current state of the philosophy of science is to understand its recent past, and that is the focus of this book.

I also go into much greater detail in describing theories, controversies, and incidents from current science and the history of science than is usual for a philosophy of science text. These narratives often go into considerable detail since I do not assume that students will have a broad background in the natural sciences or the history of science. The reason for including so many examples of actual scientific practice is that students need to be shown that the problems of the philosophy of science are *real* problems that arise from real science, and not just armchair amusements for philosophers. Also, it is important to show how real science bears on the evaluation of philosophers' claims about science. I think that one very valuable contribution of Thomas Kuhn and other "postpositivist" philosophers of science was to show that legitimate philosophy of science must be true to how science really has been done.

A detailed, chapter-by-chapter summary of this book's contents is found at the end of Chapter One. The reason it is placed there rather than in the Preface for Students is that the information given in the first chapter is really necessary for students to understand what the rest of the book is about.

In this book I have not attempted to maintain the impersonal tone of studious solemnity typical of textbooks. As is already obvious, I do not eschew the use of the first-person singular pronoun. I also write in a more conversational and informal style than is usual for textbooks. My reason for writing in this more personal idiom is simply that I find that this works best in my own teaching. Students respond much more readily to a looser conversational style than to a stiff, formal "lecture" mode. Throughout I have tried to follow Einstein's dictum that when you attempt to convey specialized knowledge to a nonspecialist audience you should explain things as simply as possible—but no simpler. So I have not shied away from or watered down some rather technical points. Though I have striven for clarity, I do not patronize students by talking down to them. I expect them to be willing to think hard and to have access to a dictionary.

Another difference between this book and the usual textbook is that I make no effort to conceal my views on controversial issues. It will be clear to any reader precisely what I think on the issues I discuss here. I do not try to hide these views behind a façade of neutrality because I think students have every right to know where the author stands on disputed points. I think that texts that present arguments in a didactic manner, with all viewpoints getting equal and neutral treatment, create a false impression—one today's students are already far too inclined to accept—that in philosophy it's all "just someone's opinion." By drawing definite conclusions and offering arguments to support them, I am driving home the point that some philosophical conclusions *are* better than others. However, I certainly do not consider any of my arguments to be the last word on any topic. One reason that I take such definite stands is, as I say above, to provide a clear focus for students and instructors to agree or disagree with me as they see fit. By stating my position and arguing for it vigorously, I am speaking to students the same way I would address a colleague, and I hope to get the same sort of thoughtful response I would expect from a colleague.

In short, it was my intention to write a text that is interesting, provocative, focused, and succinct—maybe one that students would actually *read*. This text was not written with the aim of being the only text employed in a course, nor do I ever aim to usurp the instructor's role. If, for instance, instructors find my treatment of some topics, like explanation, far too cursory, I'm sure they will have no problems supplementing my presentation.

PREFACE FOR STUDENTS ᔕ

I WON'T TRY TO GUESS what motivated you to take a course dealing with the philosophy of science. Congratulations on your adventurous spirit, though. I'm sure that "philosophy of science" sounds far too deep, difficult, and dull for most students' tastes. Practically every course in the Schedule of Classes has a sexier title than "philosophy of science." It sounds like a field full of stupefying technicalities and totally devoid of the vibrancy of human passion. Nothing could be further from the truth. Philosophy of science, like all areas of advanced research, contains some technical parts. But the big issues in the field, the ones that really matter, can be explained quite simply, in English, and without a lot of jargon or a host of arcane symbols. That is what I set out to do here. As for human passion, debates in the philosophy of science have plenty of that. In fact, during the 1990s, debates over some of the topics treated in this book got so heated that the media began to speak of the "science wars."

You may have noticed that the subtitle of this book is "A Concise Invitation to the Philosophy of Science." It is an "invitation," not an "introduction." There are quite a few good textbooks available that introduce the philosophy of science by offering an overview of the whole field. The author of such a text will try to cover all of the main topics of contemporary philosophy of science and provide a balanced overview of all different viewpoints. This book is neither comprehensive nor balanced. My aim is not to survey the philosophy of science or to present the full spectrum of opinion on all points, but to stimulate readers to think hard about a few of the most interesting and important topics in the field. In my view, the most interesting questions about science are these two: (1) In what sense, if any, is science a more "rational" or "objective" enterprise than any other intellectual endeavor? In other words, should science enjoy any special status that sets it above other "ways of knowing"? (2) Should scientific theories be regarded as true or accurate depictions of reality, or are they just useful devices for predicting and manipulating events? That is, should we trust the story science tells us, or do we accept science just because it happens to "work"? These questions will be more precisely and comprehensively phrased at the end of Chapter One.

As for presenting balanced views on each issue, some of the referees who read a draft of this book prior to publication complained that it was too opinionated. Since by purchasing this book you are the consumer of my product, I offer the following label: "WARNING: This book contains many of the author's OWN OPINIONS. These are included for educational purposes only."

My reasons for writing an opinionated text (besides the fact that it was a lot more fun to write this way) are simple. Texts that strive to give an entirely neutral, balanced, evenhanded presentation of all viewpoints create a badly misleading impression. I'm afraid the impression they create is one that many students are already overly susceptible to; namely, that when it comes to philosophical issues it's all "just someone's opinion"— that one viewpoint is really just as good as any other and it all comes down to personal preference. I think this is dead wrong. I think some philosophical opinions really are better than others because they are supported by more evidence and better argument. Therefore I unapologetically offer arguments for what I think is right. If a writer makes a claim that I think is off base, I say so. On the other hand, when somebody says something good, I give due credit.

Does the fact that I offer my own opinions mean that I am trying to bias you? No. On the contrary, I think you have the right to know exactly where the author stands on a given point so you can make up your own mind to agree or disagree. If I were trying to bias you, I would pretend to be neutral while subtly slanting things in my favor. Instead, I'm paying you the compliment of presenting you with arguments that you can think over and decide to accept or reject. One referee feared that by being opinionated, I would appear to have "an agenda." I do have an agenda— to invite you to think very hard about the deep issues I raise here. If you wind up disagreeing with any or all of my conclusions—so be it! A good teacher (which I strive to be) always prizes students who thoughtfully dissent from the instructor's own views. In fact, by stating my position and arguing vigorously for it, I am addressing you as I would a colleague. In doing so, I am inviting you to give a thoughtful response, just as I would expect from a colleague.

The topics I cover in this book are narrower than the ones found in a survey text. However, I go into these issues in much greater depth and detail than a survey text can. My experience is that students learn much more from getting into a few problems in depth than from getting a superficial overview of many topics. The writers I examine were *not* selected because they are the most representative spokespersons for their particular kind of view, but because they say controversial things and give clever arguments for their conclusions. My experience as a teacher shows me that students respond to controversy. So, I decided, for instance, to focus on a particular feminist philosopher, not because she is a *typical* feminist philosopher (whatever that might be), but because she makes a strong claim and gives a vigorous, in-your-face argument for it. You might wonder whether some of these writers really said some of the more extreme sounding claims I put in their mouths. You might wonder whether I have been entirely fair with some writers or whether I am presenting a "straw man" (or "straw woman"). I have tried hard to present the real views of the thinkers I examine, but please be sure to pay close attention to the "Further Readings" essay at the end of each chapter. There I give the sources where you can consult the works of all the people and ideas I discuss. If I attribute a view to some writer and it sounds far-fetched to you, then don't take my word for it. Read the author for yourself and see if I've gotten it right!

In writing this book I have not presumed that the reader knows much about either philosophy or science. This is part of the reason why I go into much greater detail in explaining scientific ideas and episodes from the history of science than is usual for a textbook in the philosophy of science. Another reason that I have put in so much detail about science is that I am afraid that students all too often get the idea that the problems in the philosophy of science are armchair amusements for philosophers (who presumably have too much time on their hands) and are irrelevant to real science. I tie in philosophical problems to real science to show that this is not so and that philosophical problems arise from and are relevant to actual science. Also, I have tried to use the least amount of philosophical jargon I could to explain each point. However, some terms are so useful that I have used them in various places. You may already know these terms if you have taken previous philosophy courses, but just to be sure, in this book I'll assume that you are familiar with the following terms. I'll put them right up front here instead of burying them in a glossary at the back of the book. You can either look at these now or just refer back to them when you encounter them in the text.

A PRIORI/A POSTERIORI: These two terms refer to the two different ways that a proposition (something asserted to be true) can be justified as an item of knowledge. The justification of a proposition is the rational basis for thinking it true. *A priori* propositions are justified independently of empirical evidence (see "EMPIRICAL" below). For instance, the proposition "All bachelors are unmarried males" is justified by the meanings of its words, not some empirical fact, so it is *a priori*. On the other hand, the proposition "Bachelors are, on average, less happy than married men," can only be justified by appeal to empirical evidence—a poll or study, say. Therefore it is an example of an *a posteriori* proposition.

DEDUCTION: Deductive reasoning is intended to be demonstrative. A good deductive argument, called a "valid" argument, is one in which the truth of the premises guarantees the truth of the conclusion. For instance:

> All whales are mammals.
> All mammals are air-breathers.
> Therefore, all whales are air-breathers.

Here, if the premises are true the conclusion *must* be true. Deduction is an all-or-nothing sort of reasoning. If the deduction is valid and all the premises are true, the conclusion is proven. If a deductive argument is not valid—even if the premises are all true, and the conclusion is, by accident, true—then the premises give no support to the conclusion.

EMPIRICAL: Of or pertaining to experience, including sensory experience, such as seeing or hearing as well as inner or introspective experience. Empirical knowledge is knowledge that is justified by experience or by inference from experience. Examples of empirical claims are "the butter has gone rancid," "Four out of five beer drinkers prefer Duff Beer over Fudd," or "The sun's energy is generated by the fusion of hydrogen atoms into helium." The justification—the rational basis for believing such claims—is either immediate experience or inference (sometimes at a considerable remove) from experience. Scientific knowledge is empirical knowledge because our grounds for accepting or rejecting scientific claims rest ultimately upon inferences from experience.

EPISTEMOLOGY: Epistemology is the branch of philosophy concerned with the nature, scope, and basis of knowledge. Epistemologists are especially interested in how some, but not all, of our beliefs qualify as knowledge. To count as knowledge, a belief must be true, but that is not enough. True beliefs must also be grounded, justified, or warranted; that is, there must be an adequate rational basis for holding them to be true. Epistemologists attempt to specify the necessary and sufficient conditions for beliefs to acquire such grounding, justification, or warrant.

INDUCTION: Induction is inference that is not intended to demonstrate its conclusion, only to make the conclusion more probable or plausible. Inductive reasoning is called "ampliative" because it amplifies our knowledge by adding new information not contained in the premises. In deduction, by contrast, the conclusion is not genuinely new information since the conclusion is already implicitly contained in the premises. This is an instance of an inductive argument:

> On many past occasions I have eaten oatmeal and it never
> poisoned me.
> Therefore, probably, on this occasion I can eat oatmeal without
> it poisoning me.

METAPHYSICS: Metaphysics is the branch of philosophy that inquires into the nature of reality, of what ultimately exists. Metaphysics seeks to penetrate the mere appearances of things and disclose the deep reality underlying those appearances. Questions such as whether reality is ultimately one (as "monism" maintains) or many (as "pluralism" says), or whether abstract objects like numbers or universals exist, are instances of metaphysical questions.

I found it convenient to define other terms in context as they are introduced in the text.

By the way, the hardest chapter in this book is Chapter Two. It deals with a somewhat technical but very important issue called "incommensurability." I've tried to smooth the way as much as I can, but understanding this chapter will require careful reading on your part. Once you get through that chapter, the rest should be much smoother sailing.

When you reach the end of this book, I hope you will have come to share some of my enthusiasm for debating issues in the philosophy of science. As will be obvious from the text, I am an unabashed fan of science (and so I do not much care for those who seem determined to debunk science at any cost). I think that the best science that has been done ranks with the highest achievements of humanity. This does not mean that I think science is perfect or that I have a reverential attitude towards it, as will be evident in the passages where I mention embarrassing or shameful episodes of science.

Because I think science is important, I think that the study of science by philosophers, historians, sociologists, and others is also important. In fact, I would go so far as to say that understanding the nature of science and its place in the life of our society is one of the most important tasks of the twenty-first century. If this book inspires some young scholars to dedicate themselves to that task, my efforts here will be abundantly rewarded.

ACKNOWLEDGMENTS

MUCH OF THIS BOOK WAS WRITTEN while I was on faculty development leave during the fall term of 2004. I would like to thank the faculty committee, my dean, Bruce Palmer, and the University Provost, Jim Hayes, for recommending and approving the leave time, as well as for their many encouraging remarks and expressions of support. I have always felt that my research and writing were strongly supported by administrators and colleagues, and I am grateful for that.

Jon-David Hague, Allison Rona, and the staff at McGraw-Hill were unfailingly polite and helpful during the process of producing this book. I am greatly appreciative to the following referees who read the (very raw) draft of some of the manuscript chapters and offered many helpful criticisms and comments:

- Carol Cleland—*University of Colorado*
- Christine A. James—*Valdosta State University*
- Timothy D. Lyons—*Indiana-Purdue University*
- Bonnie T. Paller—*California State University, Northridge*

I made an especial effort to rewrite sections where, as they pointed out, my rhetoric got a bit carried away. Other suggestions, such as the addition of extra topics, while eminently reasonable, could not be accommodated without fundamentally changing the nature of the text. Some criticisms of some referees struck me as wrongheaded, and I chose to ignore these. Naturally, I have nobody but myself to blame for any faults in the text resulting from this decision.

Others who read some or all of the manuscript and made many helpful suggestions include philosopher Cory Juhl of the University of Texas at Austin, philosopher Robert Almeder of Georgia State University, my niece and future historian Erin Cochran, and two of my friends here at University of Houston–Clear Lake, physicist David Garrison and biochemist Ron Mills. Everyone these days is exceedingly busy, and it means a lot to me when people take the time to read things and give thoughtful feedback. Thanks also to Kevin Rawlings for the neat computer illustrations he prepared for Chapter One.

I would like to thank my professors from the Department of History and Philosophy of Science and the Department of Philosophy at the University of Pittsburgh for creating an outstanding environment for the study of the philosophy of science. I especially would like to thank Professors John Earman and Clark Glymour who jointly conducted a seminar on the realism/antirealism issue that I attended as a grad student at Pitt. This was a terrific seminar that showed how good graduate education can (all too rarely) be. Also, I would like to take a moment to honor the memory of Professor Wesley Salmon, whose tragic death in an automobile accident saddened us all and deprived the philosophy of science of one of its most penetrating and lucid intellects.

Naturally, while writing a book you impose on your family more than anyone else. I would like to thank my wife, Carol, for putting up with the weekends spent at the office writing and revising, and for cheerfully tolerating my general air of distraction when some problem would preoccupy me. We did take one week of the faculty development leave time for a wonderful vacation in Paris. So, I lovingly dedicate this book to Carol, and I hope that whenever she sees it, she will think of Paris.

1 🖋

COPERNICAN QUESTIONS

O N MAY 24, 1543, EUROPE'S FOREMOST ASTRONOMER lay dying. The story goes that Nicholas Copernicus was on his deathbed when he received from the printer the first copies of his great work *De revolutionibus orbium caelestium—On the Revolutions of the Heavenly Spheres*. This work proposed nothing less than a radical revision of the established view of the universe. Copernicus argued that the earth is not the immovable center of the universe, but rather is part of a solar system. He proposed that the earth and its sister-planets Mercury, Venus, Mars, Jupiter, and Saturn, are arrayed in a series of concentric circular orbits about the sun (or, to be precise, a point very close to the sun). Copernicus was not the first to make this startling suggestion; several thinkers in ancient Greece had entertained the notion of a heliocentric (sun-centered) cosmology. But by Copernicus's day, the geocentric (earth-centered) cosmology had gained the full support of science, philosophy, and, most importantly of all in those days, theology. Because the geocentric view was so deeply entrenched, Copernicus's system was regarded as shocking, absurd, or perhaps even heretical. Yet despite the opposition of scientists, philosophers, and the Church—which famously condemned Galileo for defending the Copernican view—the heliocentric theory had won by about 1650.

The Copernican revolution, like the Darwinian revolution of the nineteenth century, impacted not just science, but something deep within the human psyche. Anyone who has followed the recent debates over "scientific creationism" or "intelligent design theory" knows that Darwinism continues to elicit passionate feeling. In a sense, we are also still coming to terms with Copernicus.

1

WHAT WAS COPERNICUS'S REVOLUTION?

Just what was so radical about Copernicus's theory, and why did it shock so many of his contemporaries? Why did others find it so inspiring that it is fair to say that the whole scientific revolution began with Copernicus? To answer these questions we have to get deeper into the history of science. Merely to say that with Copernicus science moved from an earth-centered to a sun-centered cosmology is hardly adequate to understand the depth and breadth of a transformation so profound that it really marks the beginning of the modern world and the demise of the medieval.

New theories do not arise in a vacuum. They are proposed in the face of established theories that have served long and honorably and have faced down many previous challengers. So, when new theories win, old theories lose. The fate of discarded theories is not pretty. They become the objects of mirth or pity as later generations find it hard to imagine how the universe could ever have been conceived in such terms.

Such condescension is inappropriate. The past is like another country, and when we disdain past views merely because they are old, we are behaving like someone who laughs at the customs and beliefs of other cultures. This does not mean that we must regard past theories—or, indeed, the customs and beliefs of other cultures—as equal to our own. The point is that theories accepted by people in previous centuries, even those theories long recognized as false, were in their day eminently reasonable views that explained the universe in a way that was rigorously logical, coherent, beautiful, satisfying, and comprehensive. The intellectual pillars of the late medieval worldview included the teachings of the Church, of course, but also the doctrines of two eminent thinkers of pagan antiquity, the philosopher/scientist Aristotle and the astronomer Claudius Ptolemy.

Aristotle (384–322 B.C.E.) is always ranked with Plato as one of the two greatest philosophical intellects of the ancient world. Aristotle studied under Plato, but he developed a distinct philosophy of extraordinary scope and power. He had important things to say about almost every conceivable topic. As opposed to Plato, whose mind ascended to the transcendent realm of eternal essences, Aristotle's intellect was focused on the physical world. The painting *The School of Athens* by the Renaissance master Raphael depicts Plato and Aristotle as its two central figures. As they walk, engaged in animated debate, Plato is pointing heavenward while Aristotle gestures with his palm down and his arm forward and parallel to the ground, as if to emphasize his concern with the earthly. It may seem odd that so worldly a thinker as Aristotle could have become the intellectual paragon of deeply religious intellectuals of Christian Europe. How this happened is one of the most remarkable stories in intellectual history.

By the late centuries of the first millennium, during the so-called Dark Ages of Europe, Aristotle's writings, except for a few of his works on logic and method, had been lost to the Latin-speaking scholars of the West. By contrast, in the East, in the intellectual centers of Muslim culture, the works of Aristotle were very well known and were the focus of much brilliant scholarship. During the twelfth century and the first half of the thirteenth, much philosophical literature, including the lost works of Aristotle, became available to scholars in the Latin West. These scholars were stunned to find in the works of a pagan philosopher a system of knowledge more comprehensive and sophisticated than anything they possessed. To get some idea of the effect, imagine what it would be like if today archaeologists unearthed tablets of ancient hieroglyphics that, when translated, contained science and philosophy more advanced than our own. Aristotle made so powerful an impression that Thomas Aquinas (1224–1274), perhaps the leading Christian thinker of the Middle Ages, referred to Aristotle simply as "The Philosopher." In *The Divine Comedy* Dante called Aristotle "The Master of Those who Know."

However, many of the more conservative elements in the Church were alarmed by the influx of Aristotelian ideas. In the year 1277 the Bishop of Paris issued an edict condemning 219 distinct propositions derived from or inspired by the philosophy of Aristotle. There were some glaring inconsistencies between what Aristotle taught and what the Church believed. For instance, Aristotle held that the world had existed forever; the Church taught, as it says in the biblical Book of Genesis, that the earth was created in six literal days. It took the genius of Aquinas and others to synthesize Aristotelian philosophy and Christian orthodoxy.

Aristotle wrote a treatise on astronomy, which was known to medieval scholars by its Latin name *De Caelo* (On the Heavens). Like all educated people of his day, Aristotle recognized that the earth is a sphere (that Christopher Columbus set out to show that the world is round is a silly but oddly persistent notion). He held that the earth stood forever as the immobile center of the universe, the hub around which everything else revolved. He also held that the cosmos is spherical, or rather a set of nested spheres centered on the earth. Aristotle endorsed the standard view that our world is composed of the four elements: earth, air, fire, and water. Further, each of these elements has a natural motion. He held that each element has weight or lightness as an innate property. Innately heavy elements like earth will tend to move in a straight line towards the center of the universe; light elements, like fire, will tend to move directly away from the center. In Aristotle's universe, up and down had absolute meanings, referring to directions towards or away from the world's center. Objects composed of the four elements can deteriorate or waste away, and the elements themselves can be changed into one another. So, the physical objects we encounter daily are unstable and impermanent; they are created and destroyed.

For Aristotle, the heavens are very different from the earth, being composed of entirely different material and obeying different laws. Each heavenly body is situated on its own sphere, and each such sphere is part of a rather complicated system of perfectly transparent crystalline spheres, each having its center at the center of the earth. The system of spheres has to be rather complicated since the movements of the heavenly bodies are complex. For instance, the sun not only makes its daily journey across the sky, but once a year it makes a complete circuit of the ecliptic, the path of the sun through the zodiacal constellations. The various rotations of the heavenly spheres, each rotating at a different uniform speed around its own axis, account for these complex motions of the heavenly bodies. The heavenly spheres are composed of a sublime substance called "aether," an indestructible fifth element. The natural motion of the aether is not up or down but to move eternally in a perfect circle. The moon occupies the lowest level of the heavenly spheres, followed by Mercury, Venus, the sun, Mars, Jupiter, Saturn, and the realm of the fixed stars (Uranus, Neptune, and Pluto were unknown to the ancients).

As you can see, there is nothing naïve or primitive about Aristotle's cosmology. It is a highly sophisticated, if premature, attempt to make systematic sense of our bewildering universe. Aristotle's system is also very beautiful, with its radiant heavenly bodies carried along on ethereal crystalline spheres. Remarkably, it is a system that is also highly congruent with common sense. Making the earth the immovable, solid center of everything certainly feels right. We still naturally speak of the sun rising and setting even though we have known for centuries that this is an optical illusion caused by the rotation of the earth on its axis. It still *looks* like we are standing still and the heavenly bodies are moving around us. For most practical purposes it is fine to think this way. Finally, and perhaps most significantly, Aristotle's cosmology fits well with Christian theology. The central place of earth, the humanly abode, is perfectly congruent with the theological view that the cosmos is the great stage where the drama of human redemption is acted out. Also, the distinction Aristotle made between the imperfect sublunary (below the moon) realm, where things decay and die, and the perfect realm of the heavens made good sense for Christians. After all, human sin has corrupted the earth, whereas the visible heavens border on the realm of God and his angels, and so approach the perfection of the divine.

Claudius Ptolemy's dates are not well known, but he seems to have done most of his work around the middle of the second century of the Common Era. Ptolemy's astronomical writings represent the culmination and synthesis of the great tradition of Greek astronomy. Yet Ptolemy was an original and creative thinker who did far more than merely compile or summarize the work of his predecessors. It is fitting that Claudius lived in Alexandria, which for several centuries had been the intellectual center of the Hellenistic and Roman world. Ptolemy's writings, like most of Aristotle's, were long lost to the West. Once again, Arabian scholars came to the rescue, compiling Ptolemy's astronomical works into a book that they simply called "*Almagest,*" which is Arabic for "The Greatest."

For Ptolemy, the aim of astronomical theory was to provide a geometrical model of the heavens that would "save the appearances." An astronomical model "saves the appearances" when its geometrical representation of the heavenly bodies accurately predicts their positions and movements. For instance, when our model accurately tells us that at a given time Mars will be in a certain place and will be observed to move in a certain way, then our model succeeds to that extent. Further, for Ptolemy and the other ancient astronomers, the mathematical model had to meet certain stringent requirements. The model had to represent the motions of planets in terms of uniform motion around perfectly circular paths. Why perfect circles? There were aesthetic and even religious reasons for insisting on perfect circles. After all, the heavens were divine, and surely the divine must move in the most beautiful way. A more "scientific" motivation for preferring circles might be that circles seemed the simplest figures. Scientists try to provide explanations that are no more complicated than they have to be. Unfortunately, as we shall see, the insistence on modeling the heavens in terms of uniform motion along perfect circles eventually led to insufferable complexity.

For Ptolemy, as for all ancient astronomers, the biggest astronomical problem was to construct a model that could account for the complex movements of the planets. "Planet" did not mean quite the same thing for Ptolemy that it does for us. "Planet" comes from a Greek word meaning "wanderer." Most of the lights you see in the night sky are the "fixed" stars, the stars that remain in the same constellations year after year. The movements of these stars are simple and regular, and they have hardly changed their relative positions over human history. If we could be transported back to the time of Ptolemy, or even to ancient Egypt or Babylon, the night sky would look very familiar to us. But the ancients recognized seven bodies—Mercury, Venus, the Moon, the Sun, Mars, Jupiter, and

Saturn—that did not follow the simple rules of the fixed stars. Ptolemy classified the sun and the moon as planets because, like the other "wanderers," they move through the constellations. Generally, they move from west to east against the background of the constellations. For instance, we might see Mars in the western part of the constellation Taurus. If we look again a week or two later we will see that it has moved considerably towards the more eastern part of the constellation. Sometimes, however, a planet will slow in its eastward journey, then stop, and then move for a while in "retrograde" fashion back towards the west. Eventually it will stop its retrograde motion and resume its normal eastward course. Planets moving in perfect circles around the earth should not show such retrograde motion. Also, the planets do not move across the sky at a uniform speed; sometimes they move faster across the constellations than at other times. How can we explain such odd and complicated movements if our model must stick to uniform motions around perfect circles?

Ptolemy's devices for solving these problems were very ingenious. One trick was the epicycle and deferent (see figure 1). Imagine a planet in a perfectly circular orbit. However, the center of the planet's orbit is not the earth or the sun but just a point in space. Suppose further that this point in space is itself moving along a circular path, a circle much bigger than the little circle of the planet's orbit. The big circle is called the *deferent* and the little circle is the *epicycle*. If you are located at the center of the deferent circle, the motion of the planet that you observe will be a compound of two different motions—the movement of the planet around the epicycle and the movement of the epicycle around the deferent. Obviously, using such devices you can make things as complicated as you like, with smaller epicycles moving around bigger epicycles and the whole system moving around a deferent. The flexibility of the epicycle-on-deferent system permits it to model very complex motions.

Two other models used by Ptolemy were the *eccentric* and the *equant*. When we say that a body is moving with uniform motion about a circle, we mean that its motion is uniform with respect to the center of the circle; that is, as viewed from the circle's center the body will sweep out equal angles in equal times. But, as noted above, the motion of planets around the earth often does not appear uniform. Ptolemy therefore placed the earth at the eccentric point, a point away from the center of the planet's orbit (see figure 2). In this way, he preserved the uniform motion of the planet—it *is* uniform around the center of its orbit—but when viewed from the eccentric point (where the earth is) it will *appear* non-uniform.

Finally, Ptolemy considered a model in which the earth and a location in space called the equant point were located at equal distances on opposite sides of the center of a planet's orbit (see figure 3). In the equant

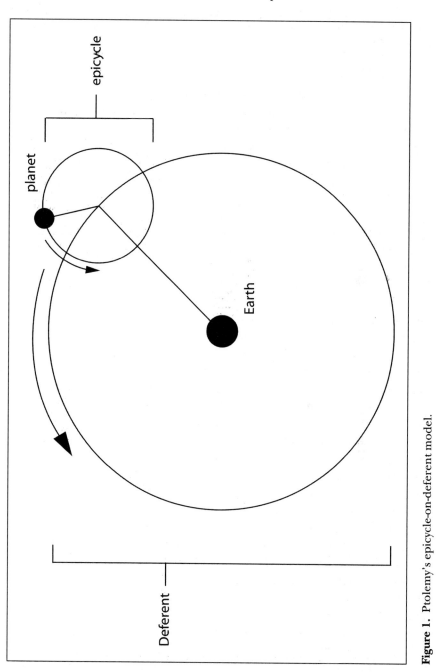

Figure 1. Ptolemy's epicycle-on-deferent model.

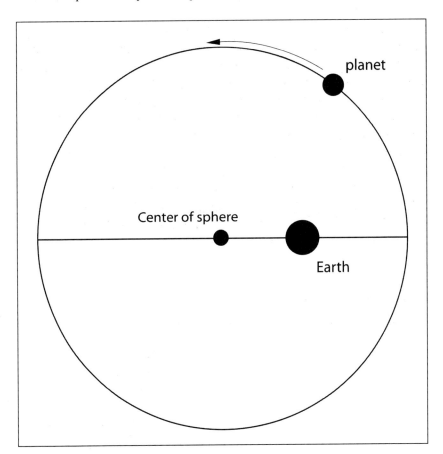

Figure 2. Ptolemy's eccentric model.

model, the planet's motion is uniform, not with respect to the earth or to the orbit's center, but with respect to the equant point. Actually, this is cheating. For the planet's motion to *appear* uniform from the equant point, the planet would *actually* have to move faster when further away from the equant point and slower when closer to it, so the equant model required Ptolemy to fudge. However, to save the appearances accurately, that is, to predict the observed motions of the planets with an adequate degree of accuracy, Ptolemy needed all three models.

Though Ptolemy's models worked fine, his system obviously got very complicated. There is a story that Alphonso X "The Wise," the King of Castille, asked his court astronomers to explain the Ptolemaic system to him. Overawed by the complexity of epicycles, deferents, eccentrics, and equants, Alphonso was heard to mutter "If the Lord had asked me for ad-

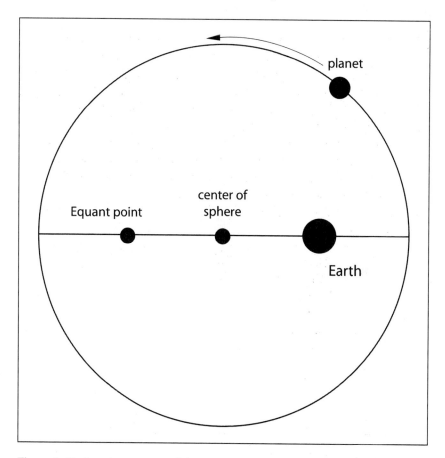

Figure 3. Ptolemy's equant model.

vice at the Creation, I would have suggested something simpler." Copernicus also yearned for something simpler. He hated Ptolemy's equant model as a violation of the rules, which in fact it was, and yearned for a system that would eliminate the equants. The result of his labors was the Copernican system (figure 4) with the sun in the center and the planets orbiting the sun in the now-familiar order: Mercury, Venus, Earth (with the Moon in orbit), Mars, Jupiter, and Saturn.

Despite the radical step of putting the sun at the center, the Copernican system was conservative in many respects. Copernicus preserved the circular orbits and the uniform motions of the geocentric astronomers. Copernicus even retained the Ptolemaic devices of the epicycle-on-deferent and the eccentric, though he did get rid of the despised equants. Still, the Copernican system in its original form was not really any simpler

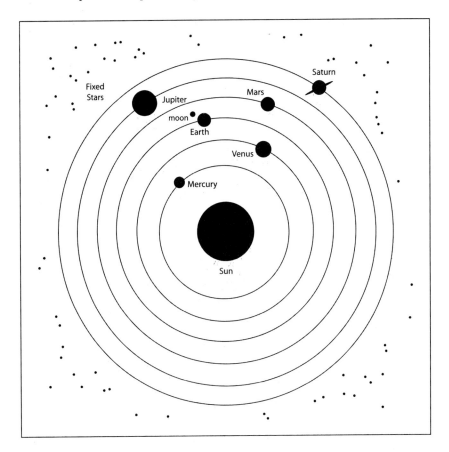

Figure 4. The Copernican system.

than the Ptolemaic system, and it could not predict the motions of the planets any better. However, Copernicus did not carry out the Copernican revolution by himself. He had the help of later astronomers who brilliantly defended and developed his system, particularly Galileo Galilei and Johannes Kepler. Kepler took the biggest step when he broke with the tradition of ages and postulated that the planets followed an elliptical rather than circular orbit. This move resulted in the great simplification of the Copernican system—no more bizarre epicycles or eccentrics—and a very considerable improvement in the accuracy of planetary predictions. The improvements in the Copernican system by Kepler, announced in his *New Astronomy* of 1609, and the brilliantly pugnacious defense of the Copernican system by Galileo in his 1632 classic *Dialogues Concerning the Two Chief World Systems,* meant that the Copernican system had its full impact only in the seventeenth century.

The acceptance of the Copernican system meant that the familiar old cosmology of Aristotle and Ptolemy had to be completely rejected. There was no place in the new cosmology for Aristotle's crystalline spheres. Worse, the telescopic observations of Galileo, published in 1610 in his *Siderius Nuncius* (The Starry Messenger), showed that the moon was not a perfectly smooth sphere as Aristotle had said all celestial bodies must be. Rather, Galileo could see that it had plains, valleys, and mountains. He could even calculate the height of some of the mountains. Old-fashioned astronomers reacted furiously, but Galileo's evidence, and his aggressive polemical style, carried the day. Kepler's elliptical orbits spelled doom for the Aristotelian notion that the movements of the heavenly bodies were "natural" and required no force to operate on them.

The final blow against Aristotle's view fell well after the Copernican view was generally accepted. In 1687 Isaac Newton published his epochal *Philosophiae Naturalis Principia Mathematica* (Mathematical Principles of Natural Philosophy). Newton added a theory of universal gravitation to the Copernican cosmology, thus showing that the same forces that govern the fall of an apple also explain the movements of the heavenly bodies. Gone forever was Aristotle's vision of a heavenly realm of absolute perfection moving in perfect spheres around the corrupt, polluted earth. Gone also was Ptolemy's geocentrism with its concept, so reassuring to theology and common sense, of a central, immobile earth. The poet John Donne (1572–1631) expressed the woe of a conservative and deeply religious intellectual for whom the "new philosophy"—the Copernican system—seemed disorienting and disturbing:

> And the new Philosophy calls all in doubt;
> The element of Fire is quite put out;
> The sun is lost, and th' earth, and no man's wit
> Can well direct him where to look for it.

Donne's perplexity is understandable. Indeed, the Copernican revolution had implications so profound that we are still dealing with them today. For one thing, after Copernicus, people had to get used to a vastly larger universe. It is not that the universe was exactly cozy prior to Copernicus. Ptolemy had regarded the earth as merely a point in relation to the whole of space. Yet the universe had to be incomprehensibly vaster for the heliocentric theory to be true. The reason is this: We can observe no stellar parallax with the naked eye. *Parallax* is the way objects in the foreground shift with respect to the background when we observe those nearer objects from different viewpoints. Close one eye and point to a distant object. Close that eye and open the other one while not moving your finger. The tip of your finger will no longer appear lined up with the distant object because the distance between your two eyes means that each eye has a slightly different perspective. The apparent shift of your fingertip with respect to the background object is parallax. If, as Copernicus claimed, the earth revolves around the sun, the stars closer to earth should shift their apparent positions with respect to more distant stars as the earth moves along its orbit. Yet the naked eye observes no such parallax in the stars (it was not observed even by telescope until the nineteenth century). The only possible explanation for the lack of visible parallax is that the stars, even the closest ones, are inconceivably far away. Copernicus was right; the stars are inconceivably far away (Alpha Centauri, the closest bright star to the sun, is 4.3 light-years or about 26,000,000,000,000—twenty-six million million—miles away).

It is not just the inconceivable distances that make us uncomfortable, it is the awareness, growing since the time of Copernicus, that our earth is not in any sense at the center of things; it is not the body around which everything else revolves. A famous poster shows a giant spiral galaxy with an arrow pointing to a minute speck in one of the outer arms. The caption reads "You are here." As the Voyager spacecraft was leaving the solar system its cameras looked back to take a final glimpse at the planets. Our earth was a tiny blue dot lost in the immensity of space. As Carl Sagan pointed out, it is humbling to think that everyone we have ever known or heard of called that tiny blue dot home. How important can we be in the whole scheme of things when our home is just one stray speck in that inconceivable immensity?

WHAT HAPPENS WHEN YOUR WORLD CHANGES?

Yet there is an even deeper sense in which the issues raised by the Copernican revolution are still with us. What happens when the effects of a scientific change are so profound that our whole worldview is altered? The Copernican revolution succeeded so completely that it is hard for us to realize how different the cosmos appears to us now than it did to even the most educated and sophisticated persons in the Middle Ages. We do not—we cannot—see the world in the same terms that our medieval ancestors did. Noted philosopher John Searle mentions a charming example. A Gothic church in Venice is called the Madonna del Orto (Madonna of the Orchard). It has that name because while it was being built, a statue of the Madonna was found in an adjacent orchard. Everyone assumed that the statue had come from heaven and signified that the church should be named after the Madonna. As Searle notes, if the church were being built today and somebody found a statue of the Madonna in the orchard, no one, not even the most devout, would assume that the statue had a heavenly origin. Everyone would just take it for granted that it had fallen there by accident or was intentionally put there by someone for an unknown purpose. Why is this? It is not that we are necessarily less religiously inclined now than we were then. Rather, we see the world in fundamentally different terms. We no longer automatically give a supernatural meaning to mysterious events. As Searle puts it: "We no longer think of odd occurrences as cases of God performing speech acts in the language of miracles. Odd occurrences are just occurrences we do not understand" (Searle, 1998, p. 35). Further, unless the occurrence is *really* odd, we just assume that it has an ordinary explanation.

Unquestionably, the main reason that we see things so differently from medieval people is that we live on the other side of the scientific revolution. The scientific revolution, of which the Copernican revolution was just the opening act, gave us a whole new set of concepts for understanding the world and imbued us with a new set of assumptions about how the world works. More profoundly, when we reject old theories and accept very different new ones, we change much more than just our beliefs. We change the way that we spontaneously act, think, and feel. It is not just that we now regard it as improbable that a Madonna statuette found in the orchard fell there from heaven. It does not even *occur* to us to think that it did, and the idea now seems quaint or superstitious to us. Put another way, when a change in outlook is sweeping enough, the transformation we experience is not merely intellectual but visceral.

The purely intellectual aspects of scientific change long preoccupied historians and philosophers of science. The historians sought to reconstruct the reasoning behind the big discoveries and to delineate the evidence from experiment and observation that convinced scientists to accept new theories. Philosophers offered formal models of scientific explanation and theory confirmation. Of course, even the most severe rationalists were aware that science was more than just an intellectual exercise, that scientists were merely human, and that scientific theories impinge on many aspects of human life. But the truly revolutionary effects of scientific change were regarded as irrelevant to the historian's or the philosopher's job, which was to focus on the rational (or, as philosophers say, the "epistemic") factors. This all changed in 1962 with the publication of Thomas S. Kuhn's *The Structure of Scientific Revolutions* (hereafter referred to as "SSR"), a work that should be on anyone's short list of the most influential books of the twentieth century.

Kuhn had a Ph.D. in physics, but he turned his attention to the history of science. In 1957 he published his book *The Copernican Revolution*, which tells the story with great clarity and insight. SSR is Kuhn's attempt to generalize the lessons he had learned by studying scientific revolutions. No summary could possibly convey the richness and subtlety of Kuhn's thought, and here I can give only a rather brief one. As Kuhn says at the very beginning of the book, he aims to show that a proper understanding of the history of science has profound implications for the image of science in our culture. Those educated in the natural sciences are taught as if their disciplines hardly had a history. From textbooks, they learn present theory and fact as though it came from no one in particular, and even if historical figures like Newton or Darwin are mentioned, only meager information is offered about them and their discoveries. By contrast, no one gets a Ph.D. in philosophy without a thorough grounding in the thought of Plato, Aristotle, Descartes, Hume, Kant, and the other major figures in the history of philosophy. A scientist generally will not care what past scientists said except insofar as their views have been absorbed, often greatly altered in form or content, into *current* theory or practice. Physics students get their Newton from current textbooks, not from reading the *Principia*. Worse, few biology students have read *The Origin of Species*.

For Kuhn, the history of science offers a number of deep lessons. First, the image of scientific change as continuous and cumulative is an illusion. True, there will be periods of what Kuhn calls "normal science" when scientists successfully solve their problems by developing and applying the guiding theories of their disciplines. During these periods scientific growth will appear stable as increments of knowledge steadily accumulate. Under the regime of normal science, a scientific discipline is committed to a core theory or set of theories that is so deeply entrenched that it largely defines the nature of the particular discipline. Kuhn calls this core theory or set of theories the *paradigm* that rules, and defines, a scientific discipline at any given time.

The role of the paradigm is crucial for several reasons. First, it defines the sorts of problems that a scientific field should study and imposes limits on the kinds of experiments or observations that will be relevant to their solution. For instance, when behaviorism was the predominant paradigm in psychology, it stipulated that behavior was what psychologists should attempt to explain, not feelings, perceptions, or other sorts of purely internal experience. Also, the data behaviorists sought were measurements of observed behavior; data from introspection were ruled out as irrelevant. Second, the paradigm delimits the kinds of answers that scientists are allowed to give when they solve puzzles. For instance, paleontologists have long been puzzled by the mass extinction that occurred at the end of the Cretaceous period, wiping out the dinosaurs and many other forms of life. No one knew what caused this mass extinction, but it was assumed that the answer, when and if it came, would involve earthly processes such as climate change or a drop in sea levels. Such earthbound causes were expected because they were consistent with the reigning paradigm in the earth sciences, which required that changes in the earth be explained in terms of the established types of slow, gradual, generally low-intensity causes that geologists had so far observed. Thus, when in 1980 a group of scientists proposed that a cataclysmic impact by an extraterrestrial body had wiped out the dinosaurs, such a hypothesis conflicted starkly with the reigning paradigm. Predictably, an enormous controversy erupted over the claim.

Things are relatively peaceful so long as a paradigm reigns unchallenged, and scientists are happily busy solving puzzles on the basis of that paradigm. Eventually, though, the paradigm runs into anomalies, facts that are difficult to accommodate on the basis of the reigning theories. Scientists will first make every effort to explain away the anomalies or show that they are, after all, compatible with the ruling paradigm. However, anomalies can pile up until a critical mass is eventually reached and, says Kuhn, science enters a "crisis" phase. In this phase, the old paradigm is effectively dead, but nothing has yet been devised to replace it. Eventually, brilliant (usually young) scientists will devise a new paradigm that resolves the anomalies that stumped the old paradigm. The new view is rapidly accepted, over the die-hard resistance of (usually elderly) scientists who cling to the old paradigm. Once the new paradigm is in place, and scientists are once more busy at puzzle solving, a new era of normal science emerges.

But what exactly happens when a new paradigm replaces an old one? The traditional view is that theory change in science is a rather pedestrian affair. New evidence, in the form of neutral observations, data, and experimental results, gradually builds up in favor of the new theory until the scientific community, calmly and rationally, decides that theory change is in order. What makes this an orderly and rational transition, on the traditional view, is that observation, data, and evidence are seen as entirely independent of the theories in dispute. In other words, the relation between theory and evidence is one-directional: Theories depend upon evidence, but evidence does not depend on theory. Therefore, evidence can serve as the objective grounds for a neutral and impartial assessment of the rival claims of competing theories.

Kuhn notes that things are not really quite so simple. A paradigm sets the standard for good science within a discipline. That is, for practitioners of that field, good science will be science like the paradigm and anything that departs too radically from the paradigm will be regarded as bad science. It follows that each paradigm comes with its own methods and standards, which are largely incompatible with those of other paradigms, and so such methods and standards cannot serve as neutral criteria for deciding *between* competing paradigms. Not even the data can be considered neutral. Each paradigm will specify what kinds of data are relevant and which can simply be dismissed. As we noted above, where strict behaviorism ruled in psychology, data about subjective feelings or inner experience were simply ruled out as irrelevant. The consequence is that scientific judgments about the value of any piece of evidence will largely be determined by the scientists' *prior* theoretical commitments. Scientists who favor paradigm A will find the types of evidence that go with A convincing; proponents of paradigm B will, on the contrary, find the types of evidence that go with B convincing. Evidence that looks good to the proponent of one theory will therefore look bad or irrelevant to the proponent of another.

In fact, a new paradigm is so different that Kuhn says that the new and the old paradigm are often "incommensurable" with each other. *Incommensurability* is a term borrowed from mathematics. For instance, the length of a side of a square is incommensurable with the length of the square's diagonal. To say that the side and the diagonal are incommensurable means that there are no possible units of any size such that whole numbers of those units will give you the length of both the side and the diagonal. Thus, there are no units that provide a "common measure" of both the side and the diagonal. Kuhn's use of the term is ambiguous; as we shall see in the next chapter he means different things by "incommensurable" at different times. However, in general, he means that proponents of rival paradigms have such different viewpoints, assumptions, and even vocabularies that they often fail to make contact when they try to communicate—they just talk past each other. When communication fails to this extent people do not even succeed in disagreeing with each other in any meaningful way; they just glare at each other across a chasm of mutual incomprehension.

So, clearly theory change in science is not the simple, straightforward process it was once thought to be. What does happen when a scientist changes his or her mind and switches from one paradigm to another? Kuhn suggests that when scientists switch allegiance from an old paradigm to a new one, this is like a religious conversion—a wholesale shift in one's view of reality. The religious convert "sees the light" or is "born again" into a whole new way of experiencing the world. Prominent religious converts such as St. Augustine or Leo Tolstoy testify that after their conversions things that previously had seemed vitally important now seemed trivial and things that had seemed silly or pointless were now full of meaning. Likewise, Kuhn suggests, the scientific convert gets a whole new perspective on reality; it is as though the convert is living in a different world than the one occupied by his or her old self. In fact, Kuhn says that the convert to a new scientific paradigm *is* living in a new world. Kuhn admits that he is not entirely sure what he means when he says, for instance, that Aristotle and Galileo lived in different worlds, but he thinks that it is true in some deep sense.

Kuhn may have been unsure just what he meant with his talk about "world changes," but many of his contemporaries thought they knew quite well what he meant. Kuhn was taken as advocating a form of what philosophers call "relativism," or more precisely, "epistemological relativism." Epistemological relativism is the position that there is no absolute, objective truth—no truth "out there" waiting to be discovered—but only "truth" relative to diverse cultures, conceptual frameworks, or worldviews. In other words truth is parochial, as are all of the standards for judging what is true, plausible, or rational. Different societies, or different groups within a society, may have very different principles for evaluating knowledge claims. For instance, when it comes to addressing some questions, one group might appeal to science and another group to the Bible. According to epistemological relativism, each group is merely articulating what is reasonable from its perspective, and neither can claim absolute authority for its standards. All we can say is that appealing to science is "right" for one group and appealing to the Bible is "right" for the other group.

Relativism is a very controversial doctrine. Some find it exhilarating. For them relativism liberates us from an oppressive, narrow rationalism that privileges a monolithic standard of truth and rationality; instead, our minds are opened to the many diverse "voices of mankind." Others think that relativism is an egregiously wrongheaded doctrine that subverts reason itself. In particular, they think that a relativistic view of science means that scientific change is irrational.

For much of his career, Kuhn complained that both enemies and would-be friends had badly misconstrued his arguments in SSR. He has especially emphasized that his conclusions were not as radical as many have thought, and he went to considerable effort to correct what he regarded as overstatements of his views by other people. Sadly, as every author soon comes to realize, once a book is published and launched into the world, it is no longer the author's possession. People will take it as they will, however loudly the author protests and reasserts his or her original intentions. All authors can do is express their views so clearly and candidly that perhaps readers of good will will not badly misconstrue them (and even then one of Murphy's Laws applies: "When you speak so clearly that no one can misunderstand you, someone will misunderstand you."). However, crucial passages in SSR are not terribly clear, and Kuhn's talk about "incommensurability," "world changes," and "conversions" makes it seem, when read literally, an assertion of some rather radical claims. So, in all fairness to Kuhn's interpreters, it is not unreasonable to read SSR as making some very bold claims.

Another problem with interpreting Kuhn is that there is hardly a claim in SSR that later writings did not qualify, moderate, or retract. So, a criticism of the early Kuhn may be deemed unfair if it does not take into account the later revisions. However, it was Kuhn the young radical that people heard, not Kuhn the middle-aged moderate, and my aim here is chiefly to present the *issues* SSR raised for its readers. Therefore, in this book I shall take the early Kuhn as canonical, especially the Kuhn of SSR. So, I now turn to what people, both friends and foes, heard Kuhn saying at the time.

It is plausible to take Kuhn as asserting that truth is relative to paradigms (and this is precisely how many have understood him). If this is Kuhn's claim, it means that judgments about the truth of any claim can only be made from *within* a paradigm, not across or between paradigms. After all, it makes no sense to say that one opinion is truer than another unless the two rival claims can be compared, and to be compared they have to be expressible in the same language. Kuhn's doctrine of incommensurability was taken as asserting that the claims of rival paradigms cannot even be expressed in a common language. That is, advocates of one paradigm do not even possess the terms that would allow them to state the claims of another paradigm. Yet if the claims made by rival paradigms are incommensurable in this strong sense—that is, they are not even expressible in the same language—then it makes no sense to say that the claims of one paradigm are truer, in any absolute sense, than the claims of another. Neither can there be any shared truths between paradigms because it makes no sense to say that one truth claim is the same as another if the two cannot even be stated in the same language. For instance, if we English speakers hear a German speaker say "*Schnee ist weiss*," and we find out that this means the same thing as "Snow is white" in English, then we can say that the German speaker is saying something we also think true. However, if we heard someone utter "Boojum glebt farkle," and no one could tell us how to express this in any language we know, we could not say whether or not this utterance affirmed or denied anything we think true. The upshot is that each paradigm incorporates its own "truth," which can neither contradict nor agree with the "truth" found in other paradigms.

The picture that emerges is a very disturbing one for anyone imbued with traditional views of science as rational and progressive. It looks like the history of science is not a march of more or less steady progress toward truth, but the successive unfolding of radical changes of perspective—a succession of paradigms, each bearing its own "truths." This means that Newton's view cannot have been truer, in any absolute sense, than Aristotle's, nor can Einstein's theories really be any truer than Newton's. Each paradigm comes with its own comprehensive set of "truths," i.e., its own "world."

COPERNICAN QUESTIONS: RATIONALITY AND REALISM

Paradoxically, therefore, it looks like the deepest question we can ask about incidents like the Copernican revolution, which for many would be the very touchstone of rational progress in science, is whether scientific change is really rational or progressive at all. Those revolutionary episodes in the history of science—when one whole way of looking at the cosmos is replaced by another, what Kuhn calls a change of paradigms—therefore raise two distinct but related questions: (a) Is it possible to make a rational comparison between rival paradigms so that the scientific decision to switch from one to the other can be based on objective reasoning and impartial evidence? Let us call this the Rationality Question. (b) Is it reasonable to say that one paradigm is closer to truth than another, so that science progresses towards truth as new paradigms replace old ones? Let us call this the Realism Question. If the answer to (a) is "no," then the traditional image of science as the very model of human rationality will have to be discarded. Instead, changes in science will look much more like changes in religious or political persuasion, and subject to the same sorts of psychological or sociological explanation. If a negative answer is given to (b), then we can no longer look upon truth as the goal of science. Put bluntly, we cannot say that science over the last 400 years has progressed any closer to truth.

In later writings, starting with the "Postscript" Kuhn added to the second edition of SSR, Kuhn has denied that it was ever his intention to portray scientific change as irrational. He has affirmed that he holds that scientists rationally choose between theories on the basis of usually recognized criteria such as accuracy, fruitfulness, scope, consistency, and simplicity. He says that all he ever meant to argue is that there can be no automatic, cut-and-dried methodology, decision procedure, or technique for determining which of two theories best exemplifies these sorts of virtues. Likewise, there is no way to compel agreement between proponents of different theories, no absolutely knock-down arguments or overwhelming evidence that will show one side to be absolutely right and the other one just plain silly. Further, Kuhn has denied that his doctrine of incommensurability rules out all or even most meaningful debate between advocates of different paradigms. He says he only claimed *partial* incommensurability between rival paradigms. (We shall examine in the next chapter some of the senses in which Kuhn says theories can be incommensurable.) Still, I think it is undeniable that for most readers, SSR raised both the Rationality Question and the Reality Question. That is, it made them query whether there really can be a rational basis for switching from one comprehensive outlook to another one. Also, it got them to ask whether the history of science really is a series of ever-closer approximations of the true picture of the universe.

It is possible to give an affirmative answer to the Rationality Question, and a negative one to the Realism Question. Even if there is some basis for rational comparison between two theories, that is not enough to show that either theory is closer to being true. Two equally false theories can differ a great deal in how they stand in regard to the evidence. For instance, suppose that Johnson committed a murder, but suspicion falls on two completely innocent persons, Smith and Jones. Upon investigation we find that Smith was in another country when the murder occurred but that Jones was seen within a mile of the murder scene about the time of the murder. Smith and Jones are equally innocent of the murder, but the evidence so far clearly favors the hypothesis that Jones did it.

In fact a number of prominent philosophers of science, such as Larry Laudan and Bas van Fraassen, reject the relativist view that rival paradigms are incommensurable, yet they do not regard successive scientific theories as moving closer to truth. That is, they give an affirmative answer to the Rationality Question and a negative one to the Realism Question. They hold that science is certainly rational, in the sense that rival theories can be compared on an objective basis. They also hold that science is progressive in certain senses. For instance, Laudan, following Kuhn's lead, takes the pragmatic view that new theories are better if they can solve conceptual problems and accommodate anomalies better than the old ones. "Realism" is the position in the philosophy of science that the goal of science is to discover the truth about the world—not only the truth about the observable parts of the world, but the truth about the deep, hidden structure of the cosmos, such as what it is ultimately made of. Further, realists hold that science has historically progressed towards truer views of the universe, culminating in our current theories, which we justifiably regard as approximately true. Antirealists such as Laudan and van Fraassen think that there are neither persuasive reasons to think that successive scientific theories converge towards truth, nor that current theories, however successful, are approximately true. So, antirealists hold that science, which they still view as an eminently worthwhile and reasonable activity, should not have the attainment of *theoretical* truth as its goal (learning the truth about *observable things* is fine for antirealists).

Interestingly, when Copernicus's theory first appeared in print, there was a dispute about whether the heliocentric system should be taken as claiming to be a true representation of the cosmos. In 1542, the year before his death, Copernicus entrusted the manuscript of his great work to his friend, the mathematician Georg Rheticus, who was to oversee its printing. However, Rheticus had just taken a new job in another city, and had to leave the uncompleted task in the hands of his friend, one Andreas Osiander. Osiander finished the job, but he also added to the text an unsigned and unauthorized preface that he himself had written. This unsigned preface, which readers would naturally think had been written by Copernicus himself, basically said that contents of the book should not be taken as true. Osiander's preface stated that the astronomer's job is solely to save the appearances by making whatever hypothetical suppositions are necessary to permit the accurate calculations of the movements of the heavenly bodies. In other words, the sole job of astronomical hypotheses is to permit astronomers to use the principles of geometry to describe the celestial motions correctly and predict them accurately. We need not regard such hypotheses as true or even probable. However, Copernicus *did* think his system was true, and not merely a handy, practical calculating device. When Rheticus saw the unauthorized preface, he was so incensed, that he sued to get the printer to remove it. The lawsuit failed and *De Revolutionibus* went into the world with its unauthorized addendum.

So the Copernican revolution has left us with some big questions and we are still grappling with them. First, how do we understand large-scale conceptual change, as occurs when science undergoes a revolution? That is, how do we understand a shift in perspective so deep, that we do not merely change our answers to certain questions, but change the entire conceptual framework in which those questions were posed? Can there be a basis for rational comparison between alternative worldviews, or are rival paradigms incommensurable? Second, even if we concede that there is some basis for rational comparison between different paradigms, is it reasonable to see science as progressing towards truth? However much better our current theories might "work," in some sense, than earlier theories, is there any reason to think that we have gotten any closer to the inner essence of reality?

The following chapters will explore these two big questions. Chapters Two and Three focus on the Rationality Question, and Chapters Four and Five on the Realism Question. Chapter Two focuses on the question of incommensurability as posed in various senses by Kuhn. Because many have seen incommensurability as implying that theories cannot be rationally compared, this seems to be Kuhn's biggest challenge to traditional scientific ideals. Chapter Three will examine arguments by two sociologists of science, two authors representing the "postmodernist" style of science critique, and feminist theorist Sandra Harding. These works, representing (but not necessarily representative of) the recent science critique of the "academic left," challenge standard views of scientific objectivity. These critics argue that science is merely a "social construct" like any other cultural artifact, and that science is not a value-neutral mode of inquiry, but is (and should be!) drenched with ideological and political agendas. Chapter Three also examines a challenge to scientific objectivity coming from the "academic right." Phillip E. Johnson, professor of law and well-known anti-evolutionary activist, argues that science has abandoned objectivity in favor of a dogmatic commitment to naturalism. Chapter Four deals with the question of progress in science, which is an aspect of the bigger question of how we should view the history of science. Some historians, strongly influenced by the social constructivist program, argue that even prototypical instances of scientific progress, like the development of the experimental method in the seventeenth century, are products of social and political agendas. Chapter Four examines and rebuts this claim. However, even if we admit that science does progress in some sense, it is a deep question whether it progresses towards *truth*. Chapter Four continues with an examination of Larry Laudan's claim that the history of science does not support the idea that science progresses towards truth. Although it is not as famous as Kuhn's SSR, Bas van Fraassen's book *The Scientific Image* (1980) has certainly had a profound effect on the philosophy of science. Van Fraassen argues that truth should not be the goal of science. He advocates an antirealist position he calls "constructive empiricism" that promotes "empirical adequacy"—saving the appearances—as the goal of scientific theory. Chapter Five examines van Fraassen's case for constructive empiricism and focuses on the important question of the goal of scientific inquiry.

FURTHER READINGS FOR CHAPTER ONE

The story of the Copernican revolution never has been and probably never will be better told than in Thomas Kuhn's *The Copernican Revolution: Planetary Astronomy in the Development of Western Thought* (New York: MJF Books, 1985). Kuhn provides a thorough review of the history of astronomy before Copernicus and a very clear and insightful exposition of Copernicus's innovations and how they were advanced by Galileo, Kepler, and others. Kuhn's knowledge is astonishing and he is eager to share it with the reader. Though debate continues on how to interpret and evaluate Kuhn's philosophical views, there is no question about his excellence as a historian of science.

Icon Books has a fine series called "Revolutions in Science." These works are succinct, highly readable, and authoritative. The series includes John Henry's *Moving Heaven and Earth: Copernicus and the Solar System* (Duxford, Cambridge: Icon Books, 2001). Henry's book can be read in an afternoon and, while not as detailed as Kuhn's classic, it tells the story with verve and lucidity.

One of the best introductions to Aristotle's thought, both as a scientist and a philosopher, is still G. E. R. Lloyd's *Aristotle: The Growth and Structure of His Thought* (Cambridge: Cambridge University Press, 1968). Lloyd also wrote two books that perhaps still constitute the best overall introduction to the science of the ancient Greeks: *Early Greek Science: Thales to Aristotle* (New York: W. W. Norton, 1970) and *Greek Science After Aristotle* (New York: W. W. Norton, 1973). Like Kuhn and Henry, Lloyd communicates deep learning without being boring or pedantic. Aristotle is a difficult thinker and a writer of technically proficient but flat and uninspired prose. The subtlety and beauty of his ideas are hard to communicate to modern audiences. Aristotle is also pretty deep; students often find his system a tough nut to crack. Lloyd's enthusiasm and skill as an expositor helps the reader to both understand and appreciate the scope and power of Aristotle's mind.

For a clear exposition of the Ptolemaic system, see Kuhn's book cited above. A very easy-to-understand introduction is in David C. Lindberg's *The Beginnings of Western Science* (Chicago: University of Chicago Press, 1992). Geocentric cosmology seems so "obviously" wrong to us today that it is tempting to view it with pity or amusement. Kuhn and Lindberg represent Ptolemy as the first-class scientist that he was, and present his system as the great intellectual achievement that it was.

John Searle is a particularly clear-headed philosopher and his book quoted in the text, *Mind, Language, and Society* (New York: Basic Books, 1998), is an especially accessible introduction to his thought. Searle states very clearly just how our worldview has changed in modern times compared to the ancient and medieval worlds.

Anyone with any interest in the philosophy of science, or the intellectual history of the twentieth century, should read Kuhn's *The Structure of Scientific Revolutions,* third edition (Chicago: University of Chicago Press, 1996). One measure of a book's importance is the quality of the opposition lined up against it. *Criticism and the Growth of Knowledge,* edited by Imre Lakatos and Alan Musgrave (Chicago: University of Chicago Press, 1970), is a compilation of some of the early critical reactions to SSR. Among the critics were Karl Popper, Imre Lakatos, and Stephen Toulmin—certainly three of the leading lights in the philosophy of science in the twentieth century. Another eminent early critic was Israel Scheffler, whose book-length critique of Kuhn, *Science and Subjectivity,* was first published in 1966. The second edition is available from Hackett Publishing Company, Indianapolis, Indiana (1982). These critics clearly took Kuhn to be saying that scientific change was not rational but was "mob psychology," as one of them put it. That is, they interpreted Kuhn as saying that there is no basis for objective comparison between competing paradigms and that only an emotional "conversion" could motivate a scientist to switch from one paradigm to another. In their view, Kuhn had debased science by injecting subjectivism and relativism into his analysis of scientific change. Their criticisms were therefore often mordant.

More recent examinations of Kuhn's work have been more balanced and less polemical. Perhaps the best monograph introducing Kuhn's work is Alexander Bird's *Thomas Kuhn* (Princeton: Princeton University Press, 2000). Two good collections of essays on Kuhn are *World Changes: Thomas Kuhn and the Nature of Science,* edited by Paul Horwich (Cambridge: MIT Press, 1993), and *Thomas Kuhn,* edited by Thomas Nickles (Cambridge: Cambridge University Press, 2003).

2

IS SCIENCE REALLY RATIONAL?

The Problem of Incommensurability

WE SAW IN THE LAST CHAPTER that Thomas Kuhn's groundbreaking study of the history of scientific revolutions raised many deep questions about the traditional view of scientific rationality. The challenge is this: When scientific change is so deep that it involves a fundamental alteration of our worldview, how can such change be rational? After all, when we say that our worldview has changed, doesn't this mean that our view of the *whole* world has changed—our understanding of the facts themselves and not just of our theories for explaining the facts? But if *everything* looks different from two different paradigms, and if our thinking is always governed by paradigms, where do we find any neutral ground to stand on so that we can judge between paradigms? How can there be any neutral body of evidence, or, indeed, any shared body of methods, standards, criteria, or values to guide our choice between competing theories? Aren't we just sealed in our own worldviews until, perhaps, an emotionally charged conversion experience knocks us into a different one?

Nevertheless, scientists often do change their theories, and they think that they are acting reasonably when they do so, even when the new theory involves a revolutionary change of perspective. When a promising new theory is proposed in science, it has to prove its mettle by taking on all rival theories. If it wins over all challengers, then it achieves the (almost) universal approval of the qualified experts. When consensus emerges and a scientific community crowns the winning theory, scientists working in that community are convinced that their theory-choice decision has been a rational one, made in the light of impartial evidence by a process of objective reasoning. Are scientists right about the reasonableness of their decisions, or are they just deluded or engaging in self-justifying rhetoric?

What, precisely, *is* the challenge that Kuhn poses to ideals of scientific rationality? As we have noted, there has been much controversy about this. For many of Kuhn's critics, the problem is that his conclusions about the incommensurability of rival paradigms mean that scientists have no logical basis for comparing opposing paradigms and making a rational decision between them. If paradigms cannot be compared in the light of neutral evidence and impartial reasoning, relativism seems to be the consequence. That is, we must conclude that there are no objective criteria for judging one paradigm rationally preferable or closer to truth than another. Each paradigm will have its own criteria for defining rationality, and each will be, by definition, "true" by its own criteria. But what exactly is incommensurability and how precisely is it supposed to make impossible the rational evaluation of rival paradigms?

The problem of incommensurability allegedly arises when two parties have such radically different views that their ability to communicate breaks down, at least to some degree. But to be a philosophically interesting idea, incommensurability has to mean more than this. Communication can break down for all sorts of reasons. Just prior to the Civil War, Southerners and Northerners could no longer have meaningful debates about the issue of slavery, or much of anything else. They hurled insults and epithets back and forth, delivered sonorous diatribes, and employed all the devices of the florid oratory of the day. But long before the first shot was fired at Fort Sumter they had ceased any meaningful exchange of ideas. However, the problem certainly did not seem to be that North and South spoke a different language or could find no common terms to express their disagreements. The problem was that feeling ran so high and opinion had become so polarized that hardly anyone was willing or even able to listen to reason anymore. Tragically, in a situation like this, when people can no longer settle their disagreements by rational, peaceful means, violence is almost inevitable.

Scientists being merely human, and prone to rancorous disputes the same as everybody else, it is not surprising that they too sometimes do not listen to their opponents. But it is no detriment to ideals of scientific rationality that scientists sometimes fail to live up to them. To threaten ideals of scientific rationality, incommensurability must mean more than that people have simply gotten too angry, stubborn, or indifferent to listen to each other. It must mean that people who have very different views—even if each is intelligent, has the best intentions, and is motivated by a sincere desire to understand the other side—simply cannot find logical, rational ways to settle all of their differences. Despite their best intentions, at certain points in their debates they fail even to disagree with each other in a meaningful way and they wind up just talking past one another.

But just how strong a thesis is incommensurability? How severe is the impairment of communication when proponents of rival theories are trying to have a rational debate, but unavoidably fail to make contact? Kuhn himself steadfastly maintains that incommensurability does not imply incomparability. In an essay titled "Commensurability, Comparability, Communicability" (CCC) written some years after SSR, Kuhn insists that incommensurability between theories is only partial and local and that many terms retain their meaning across theory change. Further, "The terms that preserve their meaning across theory change provide a sufficient basis for the discussion of differences and for comparisons relevant to theory choice" (CCC, p. 36).

Still, in many passages Kuhn argues that incommensurability means that theories might not be comparable in ways that everybody, from the seventeenth century on, has assumed that they are. So, it does sound like Kuhn is making a strong claim about the comparability of theories. To get clear on just what Kuhn's thesis is, we need to consider how philosophers of science prior to Kuhn viewed the process of theory choice. The reigning model of theory confirmation before Kuhn was called "The Hypothetico-Deductive Method" (H-D method). According to this model, scientists use a very straightforward and simple method for choosing between rival theories: A hypothesis (or theory; I'll use the terms interchangeably) is proposed. From that hypothesis, and certain other additional statements, an "observational consequence" is deduced. That is, if the hypothesis and the additional statements are all true, we can be sure that another statement must be true, a statement predicting that at a given time and place a particular observation will be made. If we look at that time and place and we observe what is predicted, we say the hypothesis is "confirmed"; if we observe something else, we say the hypothesis is "disconfirmed." When two rival hypotheses are being compared, we deduce contradictory predictions from each hypothesis. If we look and see that one prediction came true and the other did not, the hypothesis making the true prediction is confirmed and the other disconfirmed.

For example, in the early nineteenth century, there were two rival theories about the nature of light. One theory said that light is a stream of tiny particles; the other theory said that light is a wave. English scientist Thomas Young realized that these two opposing theories made different predictions. Wave phenomena can produce interference effects; particles cannot. Suppose that you have two identical waves that are exactly in phase, that is, the crest and trough of one wave is lined up with the crest and trough of the other. If these two waves intersect to form one wave, the crests and the troughs of the component waves will combine constructively to produce a new wave of double intensity. This is *constructive interference.* If two identical intersecting waves are exactly one-half wavelength out of phase, so that the crest of one lines up with the trough of the other, the two waves will cancel out each other. This is *destructive interference.* Particles, naturally, would not interfere in this way, but would just pile up. So Young deduced that in a certain experimental situation intersecting beams of light should produce interference patterns—alternating lines of constructive and destructive interference—if light is a wave, but should only produce intensified light if light is a stream of particles. He set up such an experiment and found that the overlapping beams definitely produced an interference pattern and not just intensified light. This experiment confirmed the wave theory and disconfirmed the particle theory.

The H-D method therefore permits rival theories to be compared on a point-by-point basis. The theories can be compared side by side because each theory implies observational consequences that are precisely denied by the observational consequences of the other. One theory predicts "p," and the other one predicts "not-p." Kuhn notes that for two theories to be comparable in this straightforward way—for their predictions to clash head to head—they must be able to express those contradictory predictions in the same terms. For instance, to conduct his experiment, Young had to be able to express intelligibly what he expected to see if light is a wave and the contrary expectation if light is made of particles. However, Kuhn asserts that with some rival theories it is not always possible to make them clash in this way. Sometimes the meanings of the terms used to express the observational consequences of one theory will mean something so different from those expressing the observational consequences of its rival—even if each theory uses the same words—that one theory no longer denies what the other asserts. It is like a situation where two people are eating bowls of chili. One takes a bite, breaks into a sweat, takes a quick swig of cold beer, and exclaims, "Man, is that chili hot!" The other person says, "Mine is cold. I'm going to have to heat it in the microwave." One says the chili is hot and the other says it is cold, but they are not really disagreeing because one means that it is extremely spicy and the other means that its temperature is too low.

What Kuhn wants to deny, therefore, is that theories are always point-by-point comparable by a straightforward examination of their observational consequences. So, for Kuhn rival theories are comparable in some ways but sometimes not in others. The problem is that Kuhn uses the term "incommensurable" in several different senses. Following W. H.

Newton-Smith in his fine book *The Rationality of Science* (1981), I identify
three senses of incommensurability in Kuhn's work: incommensurability
of standards, incommensurability of values, and radical meaning variance. Two rival theories may each claim to explain a given set of natural
phenomena, but differ considerably on the issue of what counts as a legitimate explanation. Proponents of these theories may therefore talk past
each other because each is assuming a very different standard about what
counts as a good explanation. One thinks that he has explained things
well and the other just does not get it. Theorists might also assume very
different values about what a good theory is supposed to accomplish. One
thinks that her theory is better because it is far simpler than its rivals; her
opponent favors a theory that is less simple but broader in scope (i.e., applies to more kinds of problems). The former scientist values simplicity
over scope, and the latter just the opposite. Because their underlying disagreement is about values and not the contents of the theories per se,
they might simply fail to communicate. Each will think the other is just
being pig-headed. Finally, two different theories may use terms that sound
the same, but their proponents fail to recognize that the meanings of the
terms differ radically between the two theories. In this case, scientists may
think that they are disagreeing when they argue with a proponent of the
opposing theory, but they are really just misunderstanding each other.

Since the purpose of this chapter is to elucidate the rational resources
available to scientists in choosing between theories, we shall consider incommensurability in each of these senses and determine what, if any,
problems are raised about the comparability of theories. We shall see, in
agreement with Kuhn, that even theories he regards as incommensurable
are rationally comparable in many ways. However, contrary to what Kuhn
says, we shall also see that even radically different theories can be compared point-by-point with respect to the evidence.

INCOMMENSURABILITY OF STANDARDS

One way that theories can be incommensurable, says Kuhn, is that they
have different standards about what constitutes good science:

> We have already seen several reasons why proponents of competing para-
> digms must fail to make complete contact with each other's viewpoints.
> Collectively these reasons have been described as the incommensurability
> of the pre- and postrevolutionary normal-science traditions, and we need
> only recapitulate them briefly here. In the first place, the proponents
> of competing paradigms will often disagree about the list of problems
> that any candidate for paradigm must resolve. Their standards or their

definitions of science are not the same. Must a theory of motion explain the cause of the attractive forces between particles of matter or may it simply note the existence of such forces? Newton's dynamics was widely rejected because, unlike both Aristotle's and Descartes's theories, it implied the latter answer to the question. When Newton's theory had been accepted, a question was therefore banished from science. That question, however, was one that general relativity may proudly claim to have solved. (SSR, p. 148)

A big rock in your hand feels heavy. The sensation of heaviness you feel is due to the force that pulls the rock towards the center of the earth. We call that force "gravity." Why does gravity exist? As we saw in the first chapter, Aristotle offered an answer for this: Rocks tend to fall down because they are composed of elements that possess an intrinsic natural motive force that impels them towards their natural place, which, for things composed mostly of earth (like rocks), is the center of the cosmos. Isaac Newton, however, when asked why gravitational attraction exists, like between a rock and the earth, famously said "*Hypotheses non fingo*" ("I frame no hypotheses"). Newton offered a mathematical law, the inverse square law of gravitation, that describes *how* rocks fall (and which encompasses all other gravitational phenomena) with great accuracy, but he declined to speculate on *why* gravitational force exists. Newton just dismissed the question of why gravitational force exists. He simply accepted its existence and set about giving a precise mathematical description of *how* gravity works. He held that his answer to the "how" question was enough for physics. You still sometimes hear people assert (wrongly) that science explains "how" but not "why." However, when Einstein devised his theory of general relativity, one of the remarkable things about that theory is that it explained, in terms of the curvature of space in the presence of massive bodies, just why gravitational attraction exists.

So, the question of why rocks fall down was at one time a meaningful scientific question, at another time dismissed as pointless speculation, and then, once again, considered an important scientific question to which Einstein gave a compelling answer. The lesson Kuhn draws is that Aristotelians and Newtonians on the one hand, and Newtonians and Einsteinians on the other, would have problems communicating because they have very different assumptions about what is a meaningful scientific question, much less what answers should be given to the questions.

Kuhn's concern is certainly legitimate. There have been times in the history of science when conflicting assumptions—over what sorts of questions can be asked or what sorts of answers can be given—have impeded communication and scientific progress. Let's look at what may, for many people, be a clearer example from another branch of science (physics is abstract and obscure for many people; the earth sciences are for them more accessible). Consider the tremendous imbroglio I mentioned in the previous chapter, the one that erupted in the 1980s when some scientists proposed a radical theory about the extinction of the dinosaurs. First a bit of background: In 1980, paleontologists, like other earth scientists, continued to accept many of the standards of good geological science laid down by Charles Lyell's 1830 classic *Principles of Geology.* Lyell opposed "catastrophism," geological theories that explained large-scale features of the earth's surface in terms of sudden catastrophes—such as massive floods—of a kind or degree never witnessed by humans. Lyell held that for geology to be a genuine science it had to explain the earth in terms of the gradual, steady operation of those geological forces we actually see operating around us. We can invoke floods to explain geological features, but they cannot be floods, like the biblical deluge of Noah, of a degree vastly greater than any floods geologists had witnessed.

By 1980 earth scientists had already come to accept that some violent geological events, such as earthquakes, floods, or volcanic eruptions, might have occurred with a degree of intensity that humans, in the very limited time we have had to observe, have never witnessed. However, there was still a considerable bias in favor of gradualism, that is, in favor of explaining large-scale earth changes, like mass extinctions, in terms of gradual processes, like climate change or the slow decline of sea levels, instead of sudden catastrophes.

Catastrophism returned to geology with a vengeance in June 1980 with the publication in the influential journal *Science* of a paper by maverick physicist Luis Alvarez, his son Walter, and their collaborators Frank Asaro and Helen Michel. This paper blatantly violated Lyell's restrictions by explaining the mass extinction at the end of the Cretaceous period— the famous "K/T" extinction that ended the dinosaurs—by hypothesizing that a massive extraterrestrial body, a comet or asteroid, had collided with the earth resulting in cataclysmic destruction. On this view, the dinosaurs, and upwards of 50 percent of all living species on the earth at the time, were wiped out in a single stupendous blast and the extreme climate changes that followed. By 1980 some rather large impacts by extraterrestrial bodies had been observed, most notably the Tunguska Event, something that caused widespread damage when it fell in a Siberian forest. But burning up even a few thousand square miles of forest is one thing, causing worldwide catastrophe and mass extinction is something else entirely. Needless to say, humans had never experienced such a cataclysm or anything close to it. Worse, the Alvarez hypothesis postulated a very sudden mass extinction, one achieved in a few months at most, rather than a gradual decline taking millions of years.

Predictably, the paper by Alvarez and his collaborators instigated a scientific controversy of almost unprecedented scope and nastiness. By 1994 over 2,500 articles had appeared in the scientific journals debating the impact hypothesis, as it came to be called. In some quarters all dignity was lost as scientists snubbed each other in public and were reduced to name-calling. I personally attended a conference on the topic of mass extinction where one invited scientist refused to attend because he would not participate in the same forum as one of his opponents. So polarized were the contending parties that William Glen, a scholar who chronicled the debates, said that the term "incommensurability" was far too weak to capture the complete breakdown in communication that had occurred. Actually, here we seem not to have had incommensurability so much as, like the South and the North prior to the Civil War, people who were just too angry to listen. Nevertheless, unquestionably one source of misunderstanding between the opposing sides was a differing set of assumptions about what kinds of answers science can give to questions like "What killed the dinosaurs?" For some, sudden catastrophes were acceptable, but for others only gradual earth-bound causes were permissible. Communication did break down, and, in part, for just the sort of reason Kuhn mentions. Still, if contending parties are not just too angry to talk, rational disagreement between contending scientists is possible even on topics like what constitutes a legitimate question or answer in science. In fact, in the past few years tempers have cooled enough for catastrophists and gradualists to seek common ground. (See, for instance, the discussion between catastrophist Dale A. Russell and gradualist Peter Dodson reprinted in my book *The Great Dinosaur Controversy*.)

Getting back to the case of Newton, in effect Newton was saying that scientists should not waste time addressing unanswerable questions about what causes gravity, but should get on with the fruitful task of using his formulas to predict phenomena and subsume them under the laws he had discovered. To scientists of Newton's day, this reasoning seemed compelling. Newton convinced them that they had no good answer to the question of what caused gravity, but they did have fantastically powerful tools, supplied by Newton, for understanding how it worked.

Scientists are an opportunistic lot, always willing to use whatever tools are available to get on with the job and leave aside, at least for the time being, questions that look too hard to answer. But this does not mean that those hard questions become literally meaningless or incomprehensible to them. Newton seemed to understand perfectly well the question about what causes gravitational attraction; he just thought that it neither had nor needed an answer. Further, Newtonians seem to have clearly understood the old Aristotelian explanations of why things fall; those answers just no longer satisfied them. To a Newtonian, to say that things fall because they have an intrinsic potency to move to the center of the cosmos is not an acceptable answer because (a) it presupposes a pre-Copernican geocentric cosmology, and (b) it does not really seem to *explain* at all. The playwright Molière satirized such "explanations" when he had one of his characters "explain" the fact that opium makes you sleepy by saying that opium has a "dormative [sleep causing] potency"—which is just a fancy way of stating the obvious fact that it makes you sleepy. Likewise, for Newtonians to say that earthly bodies have an intrinsic downward tendency just seemed to reiterate the everyday observation that they fall down. In short, Newton and his followers had good *reasons* for rejecting the old Aristotelian definitions and standards—reasons they could, and did, offer in debate with adherents of the old views.

In fact, by Newton's day proponents of the new science, like Francis Bacon and Galileo, had largely won the battles against Aristotelian science. The chief rival of Newton's gravitational theory was the mechanistic theory of the French philosopher/scientist René Descartes. Descartes defended a theory that explained all physical effects in terms of "corpuscles," minute particles that mechanically interact with one another. Descartes postulated a vortex of fine celestial matter spinning around the earth, sun, and each other heavenly body. This celestial matter pervades the cosmos and is less dense than the matter that composes massy objects such as rocks. The spinning motion of the vortex produces centrifugal force, the outward-pulling force that you feel when you tie something to a string and then spin it around. The centrifugal force in the spinning vortex pulls the fine particles of celestial matter outward and away from the earth. Air is composed largely of the fine celestial matter. When you release a rock into the air, the air below the rock will strive to rise due to its centrifugal tendency. The rock contains little celestial matter, and so will have much less of a centrifugal tendency than the particles of air. The rock will therefore tend to move downward to replace the rising air. It is the downward tendency of massy bodies to replace the upwardly tending celestial matter that we feel as the weight of those bodies.

Here, then, we have an explanation of gravity that does not postulate occult intrinsic propensities of motion, but accounts for everything in terms of mechanical principles—centrifugal force and density. The reigning paradigms for all physical explanation in Newton's day required the postulation of such mechanical models, yet Newton violated that requirement in his theory of gravity. Prior to that, Newton had been as true an adherent of the "mechanical philosophy" as anyone. Did Newton undergo something like a conversion experience that suddenly propelled him into a new paradigm? That is, did he have a sudden and comprehensive shift of allegiance from the mechanical view to a new paradigm, so that his old view now seemed silly or incomprehensible?

No. On the contrary, Newton's suspicions about the "aether" (the light celestial matter that Descartes invoked to explain gravity—entirely different from Aristotle's aether, by the way) grew slowly, and he rejected it hesitantly and in a piecemeal manner. Newton came to doubt the aether because, no matter how fine or rarefied it was supposed to be, a matter that pervaded the cosmos should offer some resistance to the planets and moons moving through it, but Newton could discern no such effect. Heavenly bodies moved precisely in elliptical orbits, sweeping out equal areas in equal times, just as his theory predicted, and exhibited no effects of the sort you would expect if they had to move through a resisting medium. Still, because Newton was so deeply imbued with the mechanical view, the idea that all effects must be transmitted though physical contact, and because some experimental results seemed to support the existence of the aether, he tried hard to hold on to the idea. Only after devising a particularly subtle and ingenious experiment that failed to indicate the effects of any such aether, and after unsuccessful attempts to modify the aether hypothesis to save it from its empirical difficulties, did he finally conclude that most of "aetherial space" must really be void. He was not suddenly "converted" to the view that no aether was required; after much experiment and deliberation he just could see no reasonable way to hold on to the idea. He rejected the aether hypothesis slowly, reluctantly, and apparently as a result of a perfectly straightforward (if brilliant) process of scientific reasoning.

The upshot is that the circumstances Kuhn cites in the quote on page 29 do not preclude meaningful and rational disagreement between advocates of competing paradigms, even on matters pertaining to the standards of science. Neither does it mean that individual scientists undergo sudden "conversions" to whole new sets of standards when they accept a new paradigm. Problems about standards, like what questions science should address or what sorts of explanations are required, seem to be topics that scientists can consider in a meaningful, rational way. Perhaps Kuhn means something different when he calls two theories "incommensurable" in the above passage. Perhaps he does not mean to assert that no rational disagreement over them is possible, only that there will exist no set methodology or foolproof technique for settling some issues. *If* this is all he meant to claim there, he is certainly right, but this weaker claim is fully compatible with the claim that disagreement between proponents of different paradigms can be settled rationally. In fact, in later writings, Kuhn admitted that meaningful, rational debate over such issues was possible, and affirmed that he only meant that no formal decision procedure or algorithm could settle such disagreements.

INCOMMENSURABILITY OF VALUES

However, Kuhn has other passages in SSR that suggest a different notion of incommensurability:

> There are other reasons, too, for the incompleteness of logical contact that characterizes paradigm debates. For example, since no paradigm ever solves all of the problems it defines, and since no two paradigms leave the same problems unsolved, debates always involve the question: Which problems is it more significant to have solved? Like the issue of competing standards, that question of values can only be answered in terms of criteria that lie outside of normal science altogether, and it is that recourse to external criteria that lie outside of normal science altogether that most obviously makes paradigm debates revolutionary. (110)

Here Kuhn is saying that not only might competing paradigms differ over standards—like what is a legitimate scientific question, or what kinds of explanations physical phenomena must have—but also in basic *values*. Each theory, *even in terms of its own standards,* will have its own successes and its own failures. So, when two theories compete, which should we value more, the successes of one theory or the successes of the other? Which is the greater liability, the failures of one theory or of its rival? Should we regard the successes of a theory as outweighing its failures? Suppose that a new theory explains in a very plausible and natural way many phenomena that the old one does not, yet it requires a physical process that just seems impossible. Which should we prefer in this situation, the new theory that explains so much more so well, or the old one that explains less but does not seem to involve a physical impossibility? Should we go with the promising new theory and hope that the seemingly impossible process will be shown possible, or stick with the old one and hope to solve its outstanding puzzles some day? Kuhn says that there is no logical way of settling such disputes, so the proponents of rival paradigms still fail to make contact. Only a fundamental change in the basic values of a science will allow a new paradigm to replace an old one. Once again, let us evaluate Kuhn's claim in the context of an actual scientific example, this one from the science of geology.

Again, cases from the history of science show that, as Kuhn says, differing assumptions about the values we expect a good theory to embody have sometimes led scientists to an impasse. Consider the theory of continental drift. Children looking at a world map or globe often notice how the eastern part of South America fits into the western part of Africa like two pieces of a giant jigsaw puzzle. This fact also impressed a young German scientist named Alfred Wegener. Wegener took a closer look and found that not only did the shorelines of South America and Africa fit, but much more importantly, the continental shelves fit even better. Intrigued by the idea that the two continents might actually have once been connected, he began to search the geological literature for other pieces of evidence. He found other fits between shores of continents now separated by ocean. Such evidence, while suggestive, did not really convince him. More convincing were mountain chains and other topographic features that are geologically identical to corresponding features on another, now separate landmass. For instance, the Sierras of Argentina could easily be a continuation of the Cape Mountains of South Africa, though the southern Atlantic Ocean now separates them. Had these continents been together at one time, these mountains would have formed a continuous chain.

Paleontology also supplied evidence. The fossils of some species are found only in the sedimentary rock in corresponding locales on separate continents. However, had those continents once been together, the present fossil sites would have formed a single continuous range for those organisms. If the continents were never connected, how could land-dwelling or freshwater organisms have traversed oceans to get to their habitats on other continents? Another puzzle had to do with ancient climates. In the Permian period, about 275 million years ago, many areas that are now tropical or subtropical were covered with glaciers, while parts of what is now northern Europe and Canada were tropical or subtropical. What could account for this? Further, actual measurements of the location of Sabine Island off the coast of Greenland between 1823 and 1907 indicated that the island had drifted westward by several meters a year!

Wegener offered a simple explanation for all these facts—the theory of continental drift. The present continents were at one time fused into a single gigantic landmass that Wegener called "Pangaea" ("All-Earth"). Over vast geological time, Pangaea broke up and different pieces drifted eventually far away from each other. If there had once been a single supercontinent, then it is perfectly reasonable that there should have been mountain ranges and habitats of organisms that were split apart as the continents separated, so they are now found on opposites sides of the ocean. If the continents do drift, it is understandable that parts that are now cold could have once been near the equator, and parts once on the equator are now in the arctic. Further, the measured drift of landmasses today is just the continuation of the drift that has occurred since Pangaea broke up.

Wegener's theory explained many phenomena that older theories could not. When Wegener proposed his drift theory, the two leading geological theories about the continents were permanentism and contractionism. Permanentism, as the name implies, held that the continents were original features of the earth, and have had pretty much their present shape and location ever since. Contractionism, on the other hand, held that the present continents were fragments of once much greater continents. Over time parts of these larger continents had broken off and subsided, forming ocean basins and isolating the current landmasses. These effects were caused by the actual contraction of the earth as it continued to cool from its original molten state. As the earth cools, its surface continues to shrink, and this causes the wrinkling, folding, and subsidence that created the earth's present surface features. Each of these theories had considerable backing and rested on much evidence. However, neither permanentism nor contractionism could easily explain the phenomena that Wegener accommodated with his drift hypothesis.

Despite its success in explaining things that the older theories could not, geologists did not rush to embrace Wegener's view. On the contrary, continental drift was not broadly accepted during Wegener's lifetime or for several decades afterwards. In fact, many professional geologists treated the drift hypothesis with contempt; one eminent authority even derided it as "geopoetry." New theories are often scorned, sometimes for good reasons and sometimes for bad. The main motivation for the geological community's rejection of Wegener's drift hypothesis seems to have been that it seemed to require a sheer impossibility. Ocean floors are composed of dense, basaltic rock. The continents are composed chiefly of less dense silicates. Wegener's theory therefore seemed to involve a geophysical impossibility—that drifting continents had somehow plowed through dense oceanic crust to reach their present positions. No known force was sufficient to push the continents around. Worse, any force that was strong enough to shove the lighter continents through dense seabed rock should literally have broken the continents to bits.

Here, then, we have the sort of case Kuhn was talking about, i.e., a difference in basic values between Wegener and his more conservative critics. Which should we value more, a new theory that in a very plausible and natural way accounts for otherwise inexplicable facts, or an old theory that, though it leaves these facts unexplained, does not entail an apparent physical impossibility? The empirical facts were not in dispute between Wegener and his critics so much as their very different assumptions about theoretical values—explanatory success versus physical plausibility in this case. These differences meant that Wegener and his critics were simply at loggerheads, and continental drift, for all its promise, was for decades relegated to the margins of geological science.

However, the subsequent history of the continental drift hypothesis shows that fundamental differences over theoretical values need not *permanently* block acceptance of a new theory, and that acceptance of that theory need not entail a "conversion" of the scientific community to a new set of values. In the 1960s a remarkable thing happened in the science of geology. At the beginning of that decade, many geologists still looked upon the drift hypothesis as heresy. By the end of the decade the consensus in the geological community (with some dissenters, of course) was that continental drift had occurred. What happened? Did the change involve a sudden shift in geologists' basic values?

Not at all. What really convinced most geologists was the development during the 1960s of the theory of plate tectonics and the startling confirmation of that hypothesis by powerful evidence. This new theory provided just what Wegener lacked, a credible mechanism for the drift of continents. According to plate tectonics, the continents rest upon massive crustal plates that themselves rest upon the earth's upper mantle. The rock of the earth's mantle is subjected to extreme heat and pressure. Under these very extreme conditions, the rock of the mantle becomes plastic and behaves in a semi-liquid manner. In fact, the earth's interior heat can cause the mantle to flow with convection currents, like a very slow-motion version of water boiling in a saucepan. In places, like in the mid-ocean ridge in the Atlantic, mantle material pushes up from deep inside the earth. As it does so, the ocean floor is pushed outwards on both sides of the ridge. The ocean floors are part of the great plates on which the continents rest, and the lighter continents float on the denser plates, so the spreading of the ocean floor can also move the continents. Startling and compelling geomagnetic evidence strongly confirmed this lateral movement of the seafloor over geological time.

The upshot is that the geological community did not have to undergo a massive shift in its basic values to accept continental drift. Rather, drift was subsumed under a well-confirmed wider theory that provided a highly credible mechanism for the migration of continents. At no point in this process did great chasms of logic or values arise between contending parties, chasms that scientists could only traverse with a leap of faith or a comprehensive conversion experience. In short, though the consequences of the emergence of plate tectonics were certainly revolutionary for the science of geology, decidedly *conservative* values and standards seemed to guide the change. Therefore, Kuhn's claim that the emergence of a new paradigm must involve a revolutionary change in scientific values does not seem to be borne out. Rather, there is far more continuity in scientific values across paradigm shifts than he admits.

INCOMMENSURABILITY OF MEANING

Still, there are other writings where Kuhn offers a more radical and more troubling notion of incommensurability. In the "Commensurability, Comparability, Communicability" (CCC) essay mentioned earlier, Kuhn argues that some terms may be genuinely untranslatable from one theory to another because we can only learn the meaning of some terms in clusters with other terms. This is because the terms in such clusters are defined in relation to each other, so that we have to grasp their meanings all together and not separately. He means something like this: Suppose that you grew up in the Amazon rainforest and that you know absolutely nothing about carpentry. At some point you see a hammer and you ask what it is. Someone tells you that it is called a "hammer," and its purpose is to drive nails. This is no help, however, because you have no idea what a nail is. Someone then shows you a nail and tells you that it is used to attach one board to another. However, you have only seen wood in its natural state, never as a finished piece of lumber, so you are still mystified. Only when someone shows you a couple of boards and demonstrates to you how the hammer and nail are used to attach one to the other do you begin to understand the purpose and nature of these strange things. So, terms like "hammer" are only really understood when we learn their meanings together with terms like "nail" and "board."

Likewise, Kuhn says that particular paradigms include sets of terms that are defined only in relation to each other in the context of that particular theory:

> In learning Newtonian mechanics, the terms "mass" and "force" must be learned together, and Newton's second law must play a role in their acquisition. One cannot, that is, learn "mass" and "force" independently and then learn empirically that force equals mass times acceleration. Nor can one first learn "mass" (or "force") and then use it to define "force" (or "mass") with the aid of the second law. Instead, all three must be learned together, parts of a whole new (but not wholly new) way of doing mechanics. (CCC, p. 44)

Newton's second law tells us how the motion of a body changes when a force is applied to it. This simple relationship is expressed by the formula $f = ma$, where f is force, m is mass, and a is acceleration. However, using the formula is one thing; really understanding the concepts "force" and "mass" in the context of Newton's mechanics is something else. Kuhn's point is that in the context of Newton's second law, you really cannot understand "force" unless you simultaneously have a grasp of "mass" (and vice versa) because the two ideas are conceptually dependent on each other. In fact they are two sides of the same conceptual coin: Force is understood in terms of how much acceleration it imparts to a given mass, and mass of a body is understood in terms of how much force is needed to impart a given amount of acceleration to that body. Now admittedly we have some understanding of what mass and force are in everyday contexts. Anyone who picks up a dumbbell can feel that it has mass, and if he drops it on his toe, he has some idea of force. But Kuhn's point is that to understand "mass" and "force" the way that *Newton* intended, we have to grasp them as interdefined in the way the second law specifies.

Suppose we try to translate the Newtonian meanings of "force" and "mass" into the language of another theory—Aristotelian or Einsteinian mechanics, for instance—that does not have Newton's second law as part of its theoretical apparatus. The translation is bound to fail, because without the second law to provide the conceptual context, "force" and "mass" just cannot be interdefined in just the way Newton meant. If you take one such term out of the context that gives it meaning and try to translate it into the language of another theory, the effort will fail because the other theory lacks the interdefined terms that give it meaning. It is easy to show mathematically that when we are considering velocities much less than the speed of light, formulas from Einstein's theory can be reduced, very nearly, to $f = ma$. However, Kuhn's point is that even if the formula is the same and the terms are the same, the meanings of "force" and "mass" are so different that the formula expresses something entirely different in Einstein's theory.

Kuhn says that even natural languages are beset with problems of incommensurability. He uses the example of the word *doux*, which, for native speakers of French, has a variety of nuances that no single term can render into English. An expert translator will use different English words or phrases to capture the different senses of such words in other languages. No single locution in English can capture all of the nuances of many terms in other languages. To cite another example, in teaching Aristotle's *Nicomachean Ethics* to students in introductory classes, it is hard to explain just what Aristotle meant by his key term *eudaimonia*. Most translators render this as "happiness," but the English word "happiness" has many connotations that do not apply to Aristotle's term, and *eudaimonia* has various senses that "happiness" just does not capture. The conclusion that Kuhn draws is that speakers of other languages, such as French or classical Greek, employ concepts that structure the world in ways that simply do not map onto the concepts employed by English speakers.

What conclusion does Kuhn want to draw from these instances of seeming incommensurability between scientific and natural languages? Apparently it is this: Where incommensurability occurs, i.e., when key concepts or clusters of concepts simply cannot be translated from one language to another, it is impossible to make a side-by-side comparison of those concepts and so determine which is more reasonable—the concepts employed in one language (or theory) or those employed in another language (or theory). For instance, if we tried to determine whether the Newtonian or Einsteinian concept of "mass" was better supported by the scientific evidence, our efforts would fail since the same word "mass" has different meanings in each theory, and those meanings cannot be translated from one theory to the other. The consequence is that even if a Newtonian and an Einsteinian appear to disagree—one says that mass is invariant through changes of motion and the other says that mass changes as bodies are accelerated—they are not really disagreeing but asserting entirely different things.

Note that for Kuhn not only do the meanings of key theoretical terms change from theory to theory, observational terms change too. For instance, it may seem like a Newtonian and an Einsteinian can make contradictory predictions about the observed mass of bodies. Suppose that a Newtonian and an Einsteinian physicist each weigh a body and agree that it has 10 kg of mass. Now the body is accelerated to 99.999% of the speed of light while the two physicists remain behind as stationary observers. What does each predict that they will observe to be the mass of the rapidly moving body? The Newtonian says that it will still be 10 kg; the Einsteinian says it will be a lot more. It looks like the two different theories therefore predict contradictory observational consequences, and simple measurements could confirm the one and disconfirm the other. However, Kuhn regards this appearance as misleading. In his view, because "mass" means something so different in each theory, the observational statement "this body has a mass of 10 kg" would have meant two incomparably different things for Newton and for Einstein, even though they used the exact same words.

It seems to follow that supporters of different theories cannot always rationally disagree with each other, because each means something radically different from the other even when they both employ the same terms like "mass" or "force." It is this claim of Kuhn's, which we may call his thesis of "radical meaning variance," that is his most troubling notion of incommensurability. If two theories are incommensurable in this sense, then it does seem that proponents of the different theories are condemned on many occasions simply to talk past one another. In this case, choosing between the two theories seemingly cannot be a straightforward rational matter, at least not in the point-by-point manner postulated by the H-D method.

EVALUATING MEANING INCOMMENSURABILITY

But do not Kuhn's own words belie this very conclusion? In various writings Kuhn has explained very clearly and in detail the differences between Aristotelian, Newtonian, and Einsteinian ideas. In fact, elucidating the differences between the different theories that reigned at given times is a very large part of what historians of science do. Scientists are probably at least as smart as historians of science, so why cannot scientists also come to grasp all of the subtleties and nuances of different theories in much the same way that historians do? Further, in the CCC essay, Kuhn himself explains very clearly—in English—the various shades of meaning of the French word *doux*. Does this not show, even though no single English word or phrase can translate the French word, that the meanings of that word are quite expressible in English? Why then cannot a physicist, using the same interpretive skills that historians of science regularly employ, grasp the meanings of two different theories and make a rational, responsible choice between the two? Kuhn's reply is that in learning the languages of two different theories, we are not translating them into the same language, but, in effect, becoming bilingual. A physicist might learn to speak both Newton-language and Einstein-language, and have an expert knowledge of both theories. But comprehending two different theories is not the same as being able to translate one into the other so that point-by-point comparisons are possible.

Now Kuhn certainly seems to be right that there are some theories that have key concepts that just do not precisely map onto the concepts of other theories, even if the two theories employ many of the same words. Let's consider another example that is probably more familiar to readers than Newtonian and Einsteinian mechanics. It is hard to think of two better candidates for incommensurable theories than Darwinism and Young Earth Creationism (YEC for short). These two different theories certainly embody two radically different and incompatible worldviews. Further, key terms in one theory just do not translate into the other. For instance, creationists often speak of "design," by which they mean the plan imposed by a Creator. However, a staunch Darwinian (and atheist) like Richard Dawkins can also speak of the "design" of an organism.

Dawkins clearly does not intend his term to imply in any sense the designing activity of a Creator. For Dawkins, "good design" in an organism is fine-tuned adaptation to a particular environment and is the result of the operation of the "blind watchmaker"—natural selection. Clearly, though they both employ the word "design," the creationist notion of design is not translatable into the language of Darwinism, since Darwinism lacks the concept of a Creator. Also, just as *doux* cannot be translated by one particular English word or phrase, so the creationist term "basic kind" cannot be rendered into any one taxonomic category in Darwinian theory. When creationists use their taxonomic term "basic kind" (after Genesis 1:24 "Let the earth bring forth the living creature after his kind . . ."), the nearest equivalent in Darwinian terminology is sometimes "species," but other times "genus," "family," or "order." So Darwinism and YEC certainly seem to be theories that Kuhn would regard as incommensurable since their key terms do not intertranslate.

Does this mean that no rational assessment of Darwinism *versus* YEC is possible, that the proponents of each theory are condemned simply to miscommunicate? Now I personally have participated, on the Darwinian side, in a number of such debates. I will testify that the experience was frustrating in various ways, and that there were indeed times that my opponents and I seemed not to be communicating. Yet it certainly does seem to me that fully rational critique of one theory by a defender of the other is possible and that powerful, indeed compelling, reasons can be given for accepting one theory and rejecting the other.

Consider a standard creationist gambit. Creationists frequently charge that evolutionary theory cannot be acceptable because it contradicts the second law of thermodynamics. The second law of thermodynamics states the principle of entropy—that in all closed systems there is a spontaneous and inevitable tendency for order to decay into disorder. They charge that evolution contradicts the principle of entropy, because evolution implies that systems automatically develop from less ordered to more ordered states. After all, evolution says that a horse evolved from a distant one-celled ancestor and a horse is obviously more complexly ordered than a one-celled organism. Now the charge that evolution violates the laws of thermodynamics is dead wrong, but the point is this: *If* evolution really were inconsistent with thermodynamics, this *would* be a major, perhaps fatal problem with evolution. The laws of thermodynamics are extremely well established, and certainly are accepted by all evolutionists. The reason why there cannot be a perpetual motion machine is that any such machine could not operate without flouting the laws of thermodynamics. If evolution were contrary to those laws, it would be just as impossible as perpetual motion machines. The point is, that though the creationist charge is in fact wrong, if it were right, evolution would be dead in the water, and evolutionists would have to admit this.

Consider now a serious problem with YEC. According to YEC, the fossil record was laid down suddenly, in the biblical flood of Noah. Why, then, are certain kinds of organisms found consistently only in the higher strata of the geological column and others consistently found much lower? One answer that creationists give is that some animals were more mobile than others and so were able to escape to higher ground before the rising floodwaters engulfed them. Thus, horses are consistently found higher than trilobites because horses would have been able to run to high ground and so escape burial much longer. In response, evolutionists have gleefully pointed out that many kinds of rooted, completely immobile plants are found only in the lower strata, and other kinds only in the upper ones. Clearly, oak trees did not uproot themselves and head for the hills as the waters rose! So for this, and many, many other reasons, the creationist account is simply incompatible with the fossil record. The upshot seems to be that, though Darwinism and YEC are candidates for incommensurability if any two theories are, the advocate of one theory can, in principle, adduce powerful, perhaps fatal criticisms of the other theory.

To make the above point stronger, let us look at some of the kinds of arguments that scientists actually use in their debates with colleagues. Our guide here is the book *The Discourses of Science* (1994) written by Marcello Pera. Pera agrees with Kuhn in rejecting the old idea of scientific method—that a single set of universal rules can prescribe foolproof procedures that automatically tell us which theories to prefer. Rationality is clearly a lot richer and messier than that. Pera argues that to understand scientific rationality we have to understand the rhetoric of scientific discourse. The term "rhetoric" now has unfortunate connotations. Political diatribes are dismissed as "mere rhetoric" by opposing politicians. "Rhetoric" in this sense means arguments based on emotional appeals and faulty logic that are used to whip up partisan sentiment and do not contribute to rational discussion. When Pera speaks of rhetoric he means it in the sense that it was originally employed when rhetoric was recognized as one of the core disciplines of the liberal arts curriculum. In this traditional sense, rhetoric is the art of constructing persuasive arguments; there is no connotation that those arguments must be fallacious or misleading.

Pera examines several works, including that masterpiece of scientific argument, Charles Darwin's *The Origin of Species*. Darwin himself called *The Origin* "one long argument" and fortified his case with many different kinds of evidence and ways of arguing. For instance, he defends natural selection by an argument from analogy. Darwin notes that everyone is familiar with the fact that by selective breeding farmers can greatly alter the nature of crops and domesticated animals. The wild turkey, for instance, is smart, elusive, and tough. The domesticated turkey, the direct descendant of the wild turkey, is stupid, docile, and delicious. The corn on the cob now enjoyed at barbecues is descended from wild maize that had an ear smaller than the baby corn now served in Chinese dishes. Clearly, plants and animals have been greatly altered to suit the needs and desires of humans. By analogy, says Darwin, nature, acting over geological ages and on all the traits of organisms, can be expected to have produced much greater effects than agriculture working over a few thousand years to alter just a few traits. Darwin also frequently argued that the best explanation of many well-known biological phenomena—well known to opponents as well as supporters of evolution—is that organisms have inherited traits, often greatly altered in form or function, from distant ancestors. For instance, there are numerous examples in creatures, including humans, of what are called "vestigial organs," structures that have no present purpose, but make sense if they are seen as remnants of organs that were useful to ancestors. For instance, people retain muscles and nerves that try to make fur stand erect when we are cold or frightened; we call this "getting goose bumps." The problem is that we lost the fur millions of years ago.

In *The Origin,* Darwin repeatedly puts his theory of natural selection head-to-head against creationism. Time and again he argues that special creation can offer no explanation for numerous phenomena that are easily explained by natural selection. Why, for instance, would a Creator, who would certainly not lack creativity, have employed the same framework of bones to create the hand of a man, the leg of a horse, the wing of a bat, and the flipper of a dolphin? Though these appendages look very different from the outside, a careful examination of the internal anatomy shows that they match up very closely, often bone for bone. These deep skeletal similarities are called "homologies." Vestigial organs, homologies, and numerous other such facts have no explanation for creationists except the inscrutable will of the Creator. Darwin argued that all these things and many more are easily explained if natural selection must work only on those features organisms have inherited from their ancestors. Note also that all these odd facts were not discovered by Darwin, but by scientists who strongly believed in a Creator.

Prima facie, these cases, and innumerable others that could be mentioned, certainly seem to show that proponents of rival theories have many legitimate argumentative strategies for attack and defense, even when the theories cannot be fully translated into a common language. Supporters of rival theories can argue, as did Darwin, that the opposing theory fails, *even on its own terms,* to account for numerous phenomena within its domain (i.e., what possible explanation could a creationist give for goose bumps?), phenomena that one's own theory easily accommodates. Scientists can argue that opponents must endorse claims incompatible with universally accepted background knowledge, like the laws of thermodynamics. These are just a few of the very many types of arguments that scientists can adduce against their rivals. Apparently then, the fact that two theories are non-intertranslatable does not at all prevent them from being compared side-by-side to see which is the more reasonable in the light of scientific evidence and logical argument. It just does not seem necessary that one theory be fully translatable into the language of another in order for there to be a meaningful comparison of one with the other or to make a completely rational choice between them. Even point-by-point comparisons seem possible. For instance, evolution by natural selection apparently explains homologies and vestigial organs and creationism does not.

Kuhn could insist that appearances are deceiving here and that Darwinism and creationism are not really compared so easily *vis-à-vis* the evidence. Recall that for Kuhn radical theory change entails not only a change in *theoretical* terms but *observational* terms also. The reason is that the vocabulary we use to describe observations is influenced by the theories we hold true (i.e., observation is "theory laden" as philosophers put it), so it might be hard for Darwinians and creationists to avoid equivocation even when discussing the supposedly neutral evidence and when using the same terms. For instance, the word "homology" means something different for creationists and Darwinians. Leading comparative anatomist Richard Owen, staunch creationist (though not of the "young earth" variety) and bitter opponent of Darwinism, carefully identified and described the homologous structures of different groups of organisms. He showed in great detail how the underlying skeletal anatomy matched up bone for bone in very different creatures. Owen held that homologies reflected what he called "archetypes"—designs in the Creator's mind that serve as generalized blueprints for organisms. For the Darwinian, homologies are not reflections of an ideal, but result from modifications by natural selection of traits inherited from ancestors. More significantly, Darwin and his followers came to define homologies in terms of their development from corresponding parts in the embryos of different organisms. Owen, however, always regarded homologies as primarily revealed in adult forms. Darwinians also rejected Owen's idea that parts within the *same* organism could hold homologous relationships with other parts. For Darwin and his supporters, homologies hold only corresponding anatomical features of different kinds of organisms.

In other words, to employ a bit of philosophical jargon, Owen and Darwin disagreed over both the "intension" and the "extension" of the term "homology." The "intension" of a term is the set conditions a thing must meet to be correctly called by that term. For instance, to be human, according to some philosophers, one must meet the joint condition of being an animal and being rational (a god is rational but not an animal; a dog is an animal but not rational). The "extension" of a term is the set of all objects referred to by the term. The set of all past, present, and future humans is therefore the extension of the term "human." For Darwin and Owen the intension of "homology" differed since they proposed different conditions for structures to qualify as homologous. Also, they differed over the extension of the term since Owen held that it could apply to different structures in the same organism and Darwin did not. So, when Darwin says that his theory explains homologies and creationism does not, does he not mean by "homology" something so different from what the creationist means that he and creationist opponents are simply equivocating rather than disagreeing?

But even if we admit that "homology" meant something entirely different for Darwin than for Owen, this objection does not wash. To confute Owen, Darwin need not claim that he means by "homology" the same thing, or anything very similar, to what Owen meant. Darwin could argue that his theory, taken as a whole—with all its theoretical and observational terms defined in *his* sense—is more satisfactory than Owen's theory as a whole. Darwin could argue that his theory postulates an intelligible and testable causal mechanism—natural selection—that offers a richer, more detailed causal account, is broader in scope, and is more consistent with the kinds of explanatory frameworks that have proven fruitful in other sciences. So, in effect, a sort of holistic "side by side" comparison of theories is possible, even if they are not translatable into a common language. Significantly, in some of his later writings Kuhn did concede that rival theories can always be compared in terms of shared theory-choice values such as simplicity, scope, accuracy, and fruitfulness.

But *did* "homology" mean something incommensurably different for Owen than it did for Darwin? Did "mass" mean something incomparably different for Newton and for Einstein? Kuhn assumes that when theories change drastically, the meanings of key theoretical and observational terms also change drastically, so drastically that, even when the new theory retains the same terms (like "mass" or "homology"), the terms of the new theory retain nothing or almost nothing of their old meanings. Also, for the doctrine of incommensurability to have any bite, the meaning of key observational terms must change completely or nearly so in all, or at least most, episodes of radical theory change. If genuine incommensurability does occur, but only very rarely, it will be hard for Kuhn and others to justify the importance they have attached to the idea.

That they are merely equivocating when they compare the observation consequences of rival theories would certainly be news to most scientists. On the contrary, the scientists who actually participate in incidents of radical theoretical transformation take it for granted that key observational consequences predicted by the new theory contradict the predictions of the old one—and if they contradict they cannot be merely equivocal. It certainly did not seem to Darwin that he meant something totally different by "homology" than Owen. Darwinians certainly thought that they were discussing many of the same anatomical parallels that Owen adduced when he talked about homologies. In other words, Darwin held that the extension of his term "homology" overlapped the extension of Owen's term. Are the scientists right in their intuitive sense that there is continuity of meaning across theory change, at least with respect to observational terms, or are they just deluded?

Why would anyone think that scientists are pervasively mistaken, so that they literally do not know what they are talking about when they engage in debates over opposing paradigms? Several commentators have noted that Kuhn and other defenders of incommensurability are assuming a particular theory about how theoretical and observational terms get their meanings. According to this theory, theoretical and observational terms acquire their entire meanings from how they are used in the context of a given theory. If this is so, then, of course, when you change theories, the meanings of all your theoretical and observational terms change too. If the theory change is radical, the meaning change is radical too. But this is a highly controversial theory of meaning that many philosophers reject. To my mind, a far more plausible theory, one supported by much empirical research in cognitive psychology, is that meanings are formed with reference to prototypes, standard exemplars of concepts. Thus, our concept of "bird" will depend largely on what kind of bird we regard as a typical example (a robin for most people in North America). In this case, I would expect that the prototypes of "massive body" would be pretty much the same for both Newton and Einstein, so that their concepts of mass would not be so different after all (alas, we have no room to explore these fascinating ideas).

Suppose though that we concede that Newton and Einstein meant something very different by "mass." What exactly do we mean by "mean" here? Recall the distinction made earlier between "intension" and "extension." Again, the intension of a term is the set of conditions something must meet to be correctly called by a term; the extension is the set of objects to which the term refers. What Kuhn seems to be saying is that the intension of a term can change radically when we shift theories. Yet two terms with very different intensions can have the same extension. Let me illustrate: The phrases "the sixteenth president of the United States" and "Mary Todd Lincoln's husband" certainly have different intensions. An entirely different set of conditions must be met to qualify as Mary Todd Lincoln's husband than are required to be the sixteenth president. However, these phrases with very different intensions have the same extension—the particular individual Abraham Lincoln. Suppose I say "The sixteenth president of the United States was over six-and-a-half feet tall" and you say "The husband of Mary Todd Lincoln was shorter than six-and-a-half feet." We are still directly disagreeing with each other even though the phrases we have used to refer to Abraham Lincoln have very different intensions. Any evidence indicating that Lincoln was over six-and-a-half feet tall would support my statement and undermine yours. Likewise, even if one theory means something different by "mass" or even "the sun" than another, as long as the sentences expressing their observational consequences have the same objects as extensions, the two theories can be directly compared. It certainly seems that Newtonians and Einsteinians would point to the same objects (like the sun) as referents of the term "object with mass."

Suppose that defenders of the "radical meaning variance" notion of incommensurability argue that the observational consequences of radically different theories do not even refer to the same objects. That is, the extensions of their terms do not overlap. In this case, where two theories do not even refer to the same objects, how can they be *rival* theories? For two views to clash, they have to be about the same thing. If someone says, "I think Muhammad Ali was the greatest boxer ever" and you say, "No, Grover Cleveland was not the only president to have a child out of wedlock," you are not disagreeing with the statement because you are not even talking about the same two people. Two theories with no referential overlap in the theoretical and observational entities they postulate cannot even provide rival accounts of what is in the world since, according to Kuhn, two such radically different theories constitute different "worlds"!

Arguments like these eventually led Kuhn to admit that incommensurability is never more than partial and that change radical enough to engender even partial incommensurability between theories is rare. But how partial is partial and how rare is rare? As philosopher and historian of science Stephen Toulmin notes, even the Copernican revolution, Kuhn's paradigm of a paradigm shift, was worked out by rational argument every step of the way:

. . . [T]he so-called "Copernican Revolution" took over a century and a half to complete, and was argued out every step of the way. The world-view that emerged at the end of this debate had—it is true—little in common with earlier pre-Copernican conceptions. Yet, however radical the resulting change in physical and astronomical *ideas* and *theories*, it was the outcome of a continuing rational discussion and it implied no comparable break in the intellectual *methods* of physics and astronomy. If the men of the sixteenth and seventeenth centuries changed their minds about the structure of the planetary system, they were not forced, motivated, or cajoled into doing so; they were given reasons for doing so. In a word, they did not have to be converted to Copernican astronomy; the arguments were there to convince them. (Toulmin, 1972, p. 105)

The fact that Kuhn in his later years would probably have agreed with everything Toulmin says shows how little was left by then of the doctrine of incommensurability.

CONVERSION: A CONCLUDING CASE STUDY

To do justice to the discussion of incommensurability, we would have to get deeply into technical discussions of theories of meaning and reference. Here we simply do not have the space to do that. Instead, let us conclude this chapter by returning to an instance of real science and follow a particular scientist through an incident of radical theory change. Perhaps the best way to test claims about incommensurability is to follow such an individual scientist through an episode of paradigm shift. Optimally, we should look at a key player whose work was integral to the theoretical transformation. If Kuhn is right about incommensurability, a scientist instrumental in bringing in a new paradigm should experience some very significant changes in his or her psychological and cognitive makeup. As Kuhn says, such an individual would have to experience something like a "conversion." But talk of "conversion" is vague, and some of Kuhn's harsher critics have understandably dismissed it as "psychobabble."

What kinds of fundamental cognitive and emotive changes would we expect to see in one who has undergone a "conversion" from one paradigm to another? Drawing upon all the senses of "incommensurability" we have discussed in this chapter, we would expect such a scientist to display radical changes in his or her scientific standards and/or theory-choice values and/or the meanings he or she assigns to theoretical and observational terms. But what if we find that, post-"conversion," the scientist in question continues to go about science in much the same way, endorses pretty much the same standards as before, accepts the same basic values as before, seems to use scientific terms in pretty much the same way as before, and, in general, evinces no massive changes in world-view? What if we find no radical conceptual hiatus, no total displacement of a whole constellation of values, standards, or meanings? In short, what if we do not find that the post-"conversion" scientist lives in a different "world"? In this case, I think we will have to reject the Kuhnian claim that even radical theory change results in incommensurability in any sense.

Let's consider the case of David Raup, certainly one of the leading paleontologists of the twentieth century. Like other paleontologists, Raup had been deeply imbued with the idea that extinction must be due to gradual processes instead of a single stupendous disaster. When the Alvarez hypothesis about the K-T extinctions came out in 1980, with its hypothesis of sudden eradication in a single cataclysm, paleontologists were generally, often vehemently, opposed. During the 1980s, though, there were some significant defections to the catastrophist side. One of the most notable of these "converts" was David Raup. Raup even wrote about his switch from old-fashioned gradualist views to a particularly radical form of catastrophism in a best-selling book—*The Nemesis Affair.*

In *The Nemesis Affair* Raup tells how he developed an exciting theory that explained mass extinctions by postulating periodic impacts of extra-terrestrial objects. The book also tells a personal story. In 1980 Raup was one of the referees asked by the editors of *Science* to review the Alvarez article that introduced the impact theory. Raup records that he was first highly skeptical of the idea and rejected the paper with harsh and rather dismissive remarks. As the controversy over the impact theory unfolded, however, Raup says that he soon had a complete change of view and became an ardent supporter of the impact hypothesis. In fact, he and colleague Jack Sepkoski went on to develop a particularly radical version of that theory—the Nemesis hypothesis. According to this hypothesis, the sun has a dim, distant companion star following a highly eccentric orbit that brings it into the solar system every twenty-six million years or so. When the companion, the Nemesis star, enters the Oort Cloud (the hypothesized region of the extreme outer solar system that is the home of comets), its gravitational effects propel millions of comets into the inner solar system. So, every twenty-six million years or so there should be evidence of mass extinctions as comets plunge into the earth.

For a reader of *The Nemesis Affair* who is familiar with Kuhn's writings on paradigm shifts, it sounds like Raup is saying that he underwent a conversion experience. In a short period of time he underwent a radical change of perspective—not just a change of opinion, but, he indicates, a visceral, comprehensive change in outlook. Like the Apostle Paul, he "saw the light" and went from being a persecutor of the new movement to one of its most outspoken defenders. If Raup did undergo a Kuhnian conversion to the impact hypothesis, we would expect the postconversion Raup to advocate very different scientific standards or values or to use terms in entirely new senses.

But a careful look at the writings Raup produced before, during, and immediately after the period of his "conversion" show no evidence of such radical disruptions. At no point does Raup begin to advocate whole new ways of doing science or evince wholesale rejection of methods and modes of reasoning he had previously employed. Though he reached radical new conclusions, the mathematical tools and scientific techniques he used to do so were not themselves radical or unlike any he had previously employed. (In fact, as William Glen observes, defenders of the impact theory often appealed to *conservative* standards.) Nor is there any reason to think that the postconversion Raup employed any terms in wholly new senses. Raup does come to understand the nature of extinction in a new way. In another book, *Extinction: Bad Genes or Bad Luck?*, Raup rejects the old idea that extinction must be understood in Darwinian terms, i.e., as the elimination of the less fit from a given environment. Rather, he proposes that the really major instances of extinction, the mass extinctions, were due to causes so cataclysmic that the fit were wiped out with the unfit. In disasters of such magnitude no Darwinian selection takes place. So, clearly, postconversion Raup understood mass extinctions differently than before. Was his later understanding of mass extinctions incommensurably different from his earlier one? Not at all. In fact, Raup takes pains to emphasize that his understanding of extinction incorporates the Darwinian view instead of displacing it. The meaning of a key term in the new paradigm is not totally different but considerably overlaps the meaning of that term in the old one. Would Kuhn reply that scientists like Raup in situations of radical theory change simply fail to notice that the terms they employ have really taken on whole new meanings? If so, then we must conclude that scientists do not understand the meaning of their own theories and must wait for historians like Kuhn to enlighten them! Such a claim appears arrogant, to say the least, and imposes a burden of proof on Kuhn and other defenders of incommensurability that they just have not met.

So, a careful examination of the case of a scientist who apparently has undergone a Kuhnian conversion if anyone has, fails to reveal any evidence supporting any of the three theses of incommensurability examined in this chapter. Of course, Kuhnians could dig in their heels and insist that the change from gradualist to catastrophist extinction theories was not a true instance of paradigm shift, and therefore that Raup's case is not a genuine counterexample. The problem here is that one of the biggest difficulties all along with talk about paradigms has been its vagueness. Just when does theory change amount to paradigm shift? How big does it have to be? Raup and other earth scientists certainly did perceive the new extinction theories as bearing assumptions that contradicted 150 years of geological tradition. Lyell's work in many ways defined the modern science of geology, and the impact hypothesis seemed to be a spectacular resurgence of catastrophism. So, if this was not a paradigm shift, it needs to be explained why not.

In conclusion, therefore, a careful examination of each of the main senses in which Kuhn says that theories may be incommensurable—with respect to standards, values, or meanings—fails to disclose any real breakdown in communication in any of these cases. True, standards, values, and meanings do sometimes change considerably from theory to theory, and the concepts of one theory often do not map onto those of a rival theory. However, on other occasions, scientists who adopt a new paradigm do not have to undergo a conversion in their basic standards, values, or concepts. Rather, continuing research shows that the new theory can satisfy conservative demands. Often, therefore, there is much more continuity in standards, values, and meanings across theories than Kuhn, or at least the early Kuhn, seems willing to admit. Even when two theories are very different we have seen that, as Kuhn claimed and contrary to the fears of his critics, in each such case scientists have an abundance of resources for the rational comparison of one theory to another. Further, even the sorts of point-by-point comparisons Kuhn says are impossible are often quite possible despite deep differences in meaning.

Perhaps Kuhn would agree that in cases of radical theory change problems of incommensurability do not always arise, but occur only in some cases. But how many would "some" be? Unless Kuhn shows otherwise, which he hasn't, we could just assume that instances of genuine incommensurability are few and far between in the history of science. Even a hard-bitten rationalist might concede that in a few cases in the history of science new theories were proposed that were so different that they could not be compared point-by-point with old ones. Even if such cases exist, we have seen, and Kuhn freely admits, that there are still many ways of rationally comparing such incommensurable theories. The lesson I draw from a study of radical theory change is that science is rational across revolutions. That is, one paradigm can be judged more reasonable than another *vis-à-vis* observational consequences and in the light of impartial reasoning. We do not have to see each paradigm as having its own "truth" or setting its own standards of rationality. So the question of relativism is not raised by paradigm shifts. Rather, just as scientists intuitively believe, empirical testing does give objective reasons for thinking that neutral evidence favors one theory over its rival.

One problem with instigating revolutions is that once started, they are hard to stop, and they may go a lot further than intended. This was the case with Kuhn. I think it is fair to say that when he published SSR in 1962 he had no idea that by the time of his death in 1996 scholars would be marching into intellectual battle under his banner. I mentioned in the Preface to Students the "science wars" that developed in the 1990s. These were bitter conflicts that arose as part of a wider "culture war" that pitted traditionalists against radicals in a battle for the soul of academe. Traditionalists defended old-fashioned ideas about the need for impartial, disinterested, and objective scholarship. The radicals, including social constructivists, postmodernists, and some (but not all) feminists, rejected these old ideals as a hypocritical sham and insisted that, since all knowledge is inherently biased and political, the ultimate aim of scholarship should be liberation of the oppressed. An icon of the radicals, one they never failed to mention when justifying their views on science, was Thomas Kuhn. They assumed that Kuhn had already exploded the myth of natural science as the exemplar of rational inquiry, leaving them only to mop up and fill in some details. We have seen in this chapter ample reason to think that this assumption is wrong, as Kuhn himself so often insisted. Nevertheless, Kuhn was the unwilling intellectual godfather of the radical science critics. In the next chapter we turn to an examination of some of the radicals who stormed the barricades of traditional philosophy of science wielding the banner of Kuhn.

FURTHER READINGS FOR CHAPTER TWO

Though outdated, Carl G. Hempel's *Philosophy of Natural Science* has never been surpassed (Englewood Cliffs, N.J.: Prentice-Hall, 1966). Hempel was one of the foremost of twentieth century philosophers of science, and his book remains a model of clear and insightful exposition. It is still the best source for beginners who want to see what philosophy of science was like prior to Kuhn. Chapters Three and Four give a very lucid statement of the Hypothetico-Deductive method.

I think the best general discussion of scientific rationality is W. H. Newton-Smith's *The Rationality of Science* (Boston: Routledge & Kegan Paul, 1981). Newton-Smith surveys the work of leading philosophers of science Karl Popper, Imre Lakatos, Thomas Kuhn, and Paul Feyerabend. He also tackles tough issues such as whether and in what sense observation should be regarded as "theory laden," or whether rival theories are ever genuinely incommensurable. The book is rather technical in places and expects the reader to keep track of many puzzling abbreviations for key terms (a bad habit of much philosophical writing). So, parts of the book would be tough going for beginning students. However, Newton-Smith provides a cogent, and mostly quite readable, statement and defense of the rationality of science in the face of various challenges.

A terrific and I think greatly underappreciated book is Richard Bernstein's *Beyond Objectivism and Relativism: Science, Hermeneutics, and Praxis* (Philadelphia: University of Pennsylvania Press, 1983). While arguing for a number of important theses about human rationality, Bernstein devotes considerable attention to the "new" philosophers of science such as Kuhn and Feyerabend. He argues, very persuasively in my view, that the true aim of these radical-sounding critiques is not to show that science is irrational, but to attack the idea, current through much of the twentieth century, that theory-choice in science was a highly rule-governed activity that obeyed rigid norms and conformed closely to formal models of confirmation. Bernstein reads Kuhn as supporting the view that theory-choice in science, while not irrational in any sense, is a much looser activity that involves a kind of reasoning that is more practical than theoretical. Bernstein draws on the distinction made by Aristotle between *theoria,* the faculty that perceives theoretical truth, and *phronesis,* the faculty of practical reasoning whereby we deliberate between various courses of action. On Bernstein's reading, Kuhn is saying that choosing between different theories in science is much more like deliberating between alternative courses of action than performing a formal inference in logic or mathematics. Bernstein's book is addressed to professional philosophers rather than students, but, after reading *Copernican Questions,* a student should be able to tackle Bernstein's sections on the philosophy of science.

Another outstanding work dealing with the rationality of science is Marcello Pera's *The Discourses of Science* (Chicago: University of Chicago Press, 1994). Many works on the rhetoric of science are debunking efforts that try to reduce scientific discourse to "mere rhetoric." Pera's book is very different. He shows, first, that formal models of scientific rationality fail to do justice to the richness of scientific debate. Further, there is no permanent, unchanging set of rules that define scientific method. As Pera observes, it would be bad if scientists did inflexibly commit to some such rigid set of methodological rules, because what could they do if *better* methods came along? He then shows, by careful case studies of the actual rhetorical practices of top scientists, that rationality and objectivity are achieved in the very process of articulating the best arguments and accumulating the best evidence in order to persuade colleagues.

As noted in the chapter, Kuhn in his later years attempted to clarify, extend, and modify his arguments in SSR. Some of his later reflections, in fact, sound like at least partial retractions of some of his more extreme-sounding claims. This is why Newton-Smith in his chapter on Kuhn says that over his career Kuhn moved from revolutionary to moderate social democrat. Several of the essays in Kuhn's *The Essential Tension* (Chicago: University of Chicago Press, 1977), particularly the essay "Objectivity, Value Judgments, and Theory Choice," present a far more balanced and nuanced view of theory choice than was given in SSR. The essay "Commensurability, Comparability, Communicability" is found in the collection of Kuhn's writings *The Road Since Structure* (Chicago: University of Chicago Press, 2000). Kuhn is quite a clear writer and many of his essays are accessible to beginning students. The main problem with understanding them is that Kuhn was a physicist, and the majority of his examples are drawn from the history of physics and may not be familiar to most philosophy students. Steven Toulmin's very insightful analysis and critique of Kuhn is found in his outstanding analysis of conceptual change *Human Understanding: The Collective Use and Evaluation of Human Concepts* (Princeton: Princeton University Press, 1972).

Isaac Newton's accomplishments dazzled his contemporaries and even today it is hard to write about him without an effusion of superlatives. Merely to describe him as a genius seriously understates the case. His was quite possibly the greatest mathematical and scientific intellect the human race has yet produced. His intellectual incandescence and personal eccentricity, bordering sometimes on psychosis (some have speculated that his bizarre quirks were due to mercury poisoning), continue to make him an irresistible subject for biographers. A recent popular biography by a fine science writer is James Gleick's *Isaac Newton* (New York: Pantheon Books, 2003). The classic introductory account of Newton's accomplishment in the context of preceding physics and astronomy is Chapter Seven of I. Bernard Cohen's *The Birth of a New Physics* (New York: W. W. Norton, 1985). The details about Newton's rejection of the mechanical aether are found in "Newton's Rejection of the Mechanical Aether: Empirical Difficulties and Guiding Assumptions," by B. J. T. Dobbs in *Scrutinizing Science*, edited by Arthur Donovan, Larry Laudan, and Rachel Laudan (Baltimore: Johns Hopkins University Press, 1988), pp. 69–83.

Alfred Wegener stated his theories of continental drift in his book *Die Ehtstehung der Kontinente und Ozeane,* published in 1912. The fourth edition from 1929 was published in English translation as *The Origins of Continents and Oceans* (London: Methuen & Co. Ltd., 1966). A good short history of the development of the drift hypothesis and the final emergence of plate tectonics is given in A. Hallam, *A Revolution in the Earth Sciences: From Continental Drift to Plate Tectonics* (Oxford: Clarendon Press, 1973). A very interesting collection of essays telling the story of the development of plate tectonics, written by the scientists involved in developing and confirming the theory, is *Plate Tectonics: An Insider's History of the Modern Theory of the Earth,* edited by Naomi Oreskes (Cambridge, Mass.: Westview Press, 2001). One thing that makes the development of plate tectonic theory so interesting is that we have here a truly revolutionary development in geology that is extremely well documented. It is therefore an excellent test case for claims that Kuhn made about scientific revolutions and theory change. For an analysis of the geological revolution in the light of Kuhn's and other theories of scientific change, see H. E. LeGrand, *Drifting Continents and Shifting Theories* (Cambridge: Cambridge University Press, 1988).

Many volumes have been written about the battles between creationists and evolutionists. This is one of those perennial battles arising from a clash of worldviews. In reading creationist literature, I have often had the creepy feeling that maybe Kuhn was right and that in some sense these writers and I do live in entirely different "worlds." My version of reality seemed hardly to match theirs at all. Upon reflection, though, I have to think that talk about adherents of different "paradigms" living in different "worlds" really makes sense only as a metaphor for extreme differences of belief. But if creationists and evolutionists do live in the same physical world, enjoy the same endowment of human intellectual and cognitive capabilities, and equally have access to the tools and methods of modern science, how do we arrive at such divergent and apparently irreconcilable conclusions? Surely, Kuhn would say, if you reject my talk about different worlds, you at least have to answer this question. Naturally, any answer to this question offered by one of the contending parties in the debate will be tendentious; that is, it will reflect the party's conviction that reason is on his or her side. Recognizing this inevitability, I offer Nicholas Humphrey's *Leaps of Faith: Science, Miracles, and the Search for Supernatural Consolation* (New York: Copernicus, 1996). Humphrey explains very clearly why intelligent and otherwise quite rational people will accept supernatural explanations in the face of overwhelming contrary scientific evidence. Further references to literature relating to the creation/evolution controversy will be given in the Further Readings essay at the end of Chapter Three.

One of the best of the many books on Darwin and the Darwinian revolution is Michael Ruse's *The Darwinian Revolution: Science Red in Tooth and Claw* (Chicago: University of Chicago Press, 1979). The source for Richard Owen's views on archetypes and homologies is Adrian Desmond's

Archetypes and Ancestors: Palaeontology in Victorian London 1850–1875 (Chicago: University of Chicago Press, 1982). Desmond's book attempts to correct the neglect and disparagement of Richard Owen's works due to the fact that he was the loser in the debates over evolution. As Desmond notes, Owen was certainly a brilliant scientist, and his reputation should be restored. However, Desmond takes far too much of a "social constructivist" approach for my taste (more on social constructivism in the next two chapters). Richard Dawkins explains how natural selection "designs" organisms in *The Blind Watchmaker* (New York: W. W. Norton, 1987).

The loud and acrimonious debate over dinosaur extinction has resulted in the publication of several popular or semi-popular works. Most of these take a strong stand and present a case either for or against the impact theory. In my opinion, the best of the pro-impact books is James Lawrence Powell's *Night Comes to the Cretaceous* (New York: W. H. Freeman and Company, 1998). The case against the impact theory is vigorously argued by Charles Officer and Jake Page in *The Great Dinosaur Extinction Controversy* (Reading, Mass.: Helix Books, 1996). Both books are fun to read, and you learn a lot whichever view you adopt. J. David Archibald's *Dinosaur Extinction and the End of an Era* (New York: Columbia University Press, 1996) presents a more balanced view that interprets the end-Cretaceous mass extinctions as due both to gradualistic factors, like marine regression and habitat fragmentation, as well as catastrophic occurrences, like meteor impacts and massive volcanism. A serious book that traces the history of the K/T extinction debates, and presents analyses by historians, sociologists, philosophers, and scientists, is *The Mass Extinction Debates: How Science Works in a Crisis* (Stanford: Stanford University Press, 1994). The editor, William Glen, also contributed two very stimulating chapters to the book. He examines the controversy in the light of Kuhn's analysis of scientific change. Even the subtitle of the book reflects Kuhn's examination of how science operates in the "crisis" phase when an old paradigm is dying and a new one being born. Anyone wanting a succinct and balanced overview of the extinction debates should consult Chapter Eight of my book *The Great Dinosaur Controversy: A Guide to the Debates* (Santa Barbara, Calif.: ABC-Clio, 2003), pp. 121–143.

Not all outstanding scientists are outstanding authors of science written for a nonspecialist audience. David Raup has enormous respect among paleontologists and also writes very readable and interesting popular science. His book *The Nemesis Affair* (New York: W. W. Norton, 1986), tells the story of how he and a colleague developed the "Nemesis" hypothesis and the controversy that resulted. The Nemesis hypothesis is no longer discussed; scientists were deadlocked on whether the statistical evidence for periodic mass extinctions was credible. Still, the book remains fascinating reading because it tells the inside story of a scientist apparently undergoing a Kuhnian-type "conversion" to a new paradigm. I analyze this "conversion" extensively in my book *Drawing Out Leviathan: Dinosaurs and the Science Wars* (Bloomington, Ind.: Indiana University Press, 2001). Raup views mass extinction as more a matter of bad luck (cataclysmic events that wipe out whole classes of organisms) than bad genes (the traditional Darwinian view that the genetically fittest survive).

3

A WALK ON THE WILD SIDE
Social Constructivism, Postmodernism, Feminism, and That Old-Time Religion

THE CONSTRUCTIVIST CHALLENGE

IF THERE EVER HAS BEEN a "hero of science," it was Louis Pasteur. He is famous all over the world, commemorated on every milk carton with the word "pasteurized." In his native France Pasteur was honored almost as a living saint during his lifetime. Now, well more than a century after his death, the Pasteur Institute in Paris remains a leading center for biomedical research. Though he had many brilliant accomplishments, he was most honored for developing vaccines, especially the vaccine against the dreadful disease rabies. Rabies had never been a great killer, like smallpox or cholera, but fear of the disease had always been far out of proportion to the number of its victims. Paul De Kruif's classic *The Microbe Hunters* contains an unabashedly heroic account of Pasteur's conquest of rabies. Here is De Kruif's description of Pasteur fearlessly risking his own life to combat disease:

> And now Pasteur began—God knows why—to stick little hollow glass tubes into the gaping mouths of dogs writhing mad with rabies. While two servants pried apart and held open the jowls of a powerful bulldog, Pasteur stuck his beard within a couple of inches of those fangs whose snap meant the worst of deaths, and, sprinkled sometimes with a maybe fatal spray, he sucked up the froth into his tube—to get a specimen in which to hunt for the microbe of hydrophobia. (De Kruif, 1926, 169)

After a long series of excruciatingly difficult experiments, Pasteur found a way to weaken the infectious agent of rabies, now known to be a virus. He concocted a vaccine that was administered in a series of fourteen injections, starting with the most weakened virus and proceeding to the most virulent. The vaccine worked perfectly on dogs, but would it work on humans? Pasteur was reluctant to try, but the story of how he did is one of the most famous in the history of medicine. In July 1885, Joseph Meister, a nine-year-old boy from the Alsace region, was mauled by a mad dog. His mother brought him to Paris and begged Pasteur to save her child. Moved by the plight of the terrified boy and his desperate mother, Pasteur agreed to try. De Kruif tells the story:

> And that night of July 6, 1885, they made the first injection of the weak-ened microbes of hydrophobia into a human being. Then, day after day, the boy Meister went without a hitch through his fourteen injections—which were only slight pricks of the hypodermic needle into his skin. . . . And the boy went home to Alsace and never had a sign of that dreadful disease. (179–180)

Pasteur's greatest triumph occurred when nineteen Russian peasants who had been bitten by a rabid wolf nearly three weeks before were brought for his treatment. So long a time had passed since they had been attacked that few believed that Pasteur could save them. He took a ter-rible risk in trying; had he failed his vaccine would have been blamed. Usually eight out of ten people bitten by rabid wolves got rabies, and once the disease strikes death is inevitable. De Kruif records the result:

> And at last a great shout of pride went up for this man Pasteur, went up from the Parisians, and all of France and all the world raised a paean of thanks to him—for the vaccine marvelously saved all but three of the doomed peasants. . . . And the Tsar of All the Russias sent Pasteur the dia-mond cross of Ste. Anne, and a hundred thousand francs to start the building of that house of microbe hunters in the Rue Dutot in Paris—that laboratory now called the Institut Pasteur. (181)

De Kruif is unabashed in his hero worship, and the same sort of awed gratitude has been expressed by many of Pasteur's biographers. Then, in 1988, Bruno Latour—a Frenchman, no less—decided to take Pasteur down a few pegs. Latour's book *The Pasteurization of France* is anything but hero worship. In fact, it is a direct assault upon the whole notion of scientist-as-hero. The Pasteur that emerges from Latour's work is not exactly a ras-cal, but he is certainly an opportunist and a grandstanding self-promoter whose successes were more theater than science.

Why does Latour want to expose Pasteur as a clay-footed giant? La-tour is not motivated by envy or mean-spiritedness; he just does not see science as the noble, selfless, pursuit of truth carried on by a few "great men" as De Kruif and other popular writers have depicted it. For Latour science is war. Scientists may give lip service to the ideals of method and objectivity, but, just as the chaos of battle nullifies the generals' beautiful plans, so scientific battles make nonsense of such fine talk. For Latour, preaching standards and values in the middle of a scientific squabble would be like reciting the Ten Commandments in a barroom brawl.

Latour has a point. Anyone who thinks of scientists as serene truth-seekers or emotionless Mr. Spock types has another thing coming. We have already mentioned in previous chapters the vicious fight that broke out over the impact theory of mass extinction. Scientists sometimes harbor personal animosities that border on mania, and pursue vendettas with such tenacity that they harm science itself. The feud between Edward Drinker Cope and Othniel Charles Marsh, the two leading American paleontologists of the nineteenth century, is a case in point. They hated each other with a reckless intensity that tarnished their reputations and corrupted their science. According to reliable reports, Marsh ordered his workers to destroy fossil specimens rather than have them fall into Cope's hands. Cope launched a yellow-press newspaper attack on Marsh, leading to highly public mudslinging that dishonored both parties. In an effort to better his highly prolific rival, Marsh often rushed his findings into print, leading to errors that took the better part of a century to sort out.

The rancor between Marsh and Cope may have been exceptionally bitter, but in the history of science there has been no lack of conflict. Also, there is no question that some of the greatest scientists have been involved in some of the loudest disputes. But hardly any great achievement in any field has ever been accomplished without bitter, intransigent, and sometimes violent opposition. So, is science really less noble, or scientists any less worthy, because controversy always accompanies discovery? Latour puts his case this way:

> We would like science to be free of war and politics. At least, we would like to make decisions other than through compromise, drift, and uncertainty. We would like to feel that somewhere, in addition to the chaotic confusion of power relations, there are rational relations. . . . Surrounded by violence and disputation, we would like to see clearings—whether isolated or connected—from which would emerge incontrovertible, effective actions. To this end we have created, in a single movement, politics on one side and science or technoscience on the other. The Enlightenment is about extending these clearings until they cover the world. . . . Few people still believe in such an Enlightenment, for at least one reason. Within these clearings we have seen developing the whole arsenal of argumentation, violence, and politics. Instead of diminishing, this arsenal has been vastly enlarged. Wars of science, coming on top of wars of religion, are now the rage. (Latour, 1988, p. 5)

The dream of the leading thinkers of the European Enlightenment of the eighteenth century, a dream inspired by the enormous achievements of modern science as epitomized by Newton, was that the rise of modern science had, at long last, brought truly objective knowledge and the one sure method for discovery. In the minds of Enlightenment thinkers, science had ushered in the Age of Reason that would displace the ages of dogma and superstition that had gone before. Humanity could finally outgrow the endless and divisive theological disputes, and the concomitant persecutions, inquisitions, and holy wars. Freed from the ever-finer hairsplitting of metaphysical speculation, the finest human minds could now turn to the production of useful knowledge. Some of the founders of modern science, like Francis Bacon and René Descartes, believed that the discovery of the scientific method meant that such pointless controversy could end. Once we have the true method, our disagreements no longer will lead to bickering; rather, we will simply calculate. Scientific questions will be solved by appeal to universally accepted procedures and will be as calm, dispassionate, and as certain as doing sums in arithmetic. It hasn't worked out like this. On the contrary, Latour argues, in science things are settled by rhetoric, negotiation, power politics, wheeling and dealing, grandstanding, *ad hominem* attacks, and intimidation, the same as everywhere else. The most successful scientists were those who were best at forming powerful alliances, appropriating grants or scarce resources, browbeating opponents, or propagandizing their views.

There is considerable truth in Latour's gruesome depiction of scientific warfare. Scientists can play political hardball. Some, like Cope and Marsh, stoop to character assassination to deal with scientific opponents. The old saying about who you know being more important that what you know does often apply in science. Scientists are merely human and they are subject to all the weaknesses and foibles—egotism, petty jealousies, spite, and narrow-mindedness—that prey upon everyone else. Also, the hope that humanity has found a single, universal, scientific method is a pipe dream. Instead, there are many different methods for many different scientific disciplines; geology and particle physics just cannot be done the same way. In addition, within each discipline methods are changing and developing all the time. But surely Latour's view is too cynical, most would still say. Science has made some incontrovertible discoveries that have revealed much about the components and workings of nature, e.g., the blood circulates, DNA is the genetic material and it has a double helix structure, the solar system is part of a giant spiral galaxy we call the Milky Way, things are made of atoms, which are themselves composed of even smaller parts called electrons and quarks.

Latour is not impressed by such litanies of scientific achievement. In 1979 he coauthored, with Steve Woolgar, a book titled *Laboratory Life: The Construction of Scientific Facts.* Latour prepared to write this book by taking a menial job at the Salk Institute, a leading laboratory for biomedical research. This job gave him the opportunity to observe scientists at work in their native habitat, like an anthropologist who lives with a rainforest tribe to observe their customs and practices. In particular, he followed scientists through the laborious, tortuous process of trying to identify and isolate a highly elusive bodily substance called TRF for short. Latour charted the complex discussion and debate as scientists initially proposed the existence of TRF, encountered skepticism and opposition, responded to criticisms, engaged in a series of rebuttals and rejoinders, and finally succeeded in convincing their colleagues that TRF is real. Like a good anthropologist, Latour studied his subjects without accepting their worldview. Just as the anthropologist does not accept at face value the tribal shaman's claims about gods and magic spells, so Latour did not take for granted the truth of scientific claims.

In *Laboratory Life,* however, Latour and Woolgar do not simply give a detached anthropological description of the customs and beliefs of the scientific tribe. Rather they offer an analysis and interpretation that radically undercuts the claims of science to discover objective facts about the natural world. Put bluntly, their aim seems to be to debunk science. For Latour and Woolgar, scientific "facts" are not discovered; they are constructed. According to this "social constructivist" view, the so-called facts of science are mere artifacts of scientific culture, just as beliefs in gods, demons, and magical powers are artifacts of tribal cultures. According to *Laboratory Life,* scientists are in the business of generating fact-statements, but nature—conceived as something that exists "out there" independently of our concepts—has virtually nothing to do with the generation of such fact-statements. Such fact-statements emerge when a given scientific community reaches consensus on the issue, and consensus is a *rhetorical* achievement (rhetorical in the sense of "mere rhetoric," where the goal is persuasion by any means necessary). In other words, all of the methods and techniques deployed in scientific debate are really just elaborate rhetorical devices, not, as scientists like to think, reliable means of testing theory against empirical reality. Consensus emerges, and so new "facts" are established, when some group of scientists employs such rhetorical devices skillfully enough to convince, or at least silence, all opponents.

For Latour and Woolgar, the means whereby scientists generate fact-statements is really quite insidious. Every new "fact" begins as an innocent hypothesis. Everyone admits its tentative and speculative nature. But as the debate proceeds, proponents of the new "fact" use all the rhetorical means at their disposal to get skeptics to drop their opposition and accept the "fact." Once this has occurred, once consensus emerges in a scientific community, a curious process that Latour and Woolgar call "inversion" allegedly takes place. An "inversion" supposedly occurs when a scientific community forgets that its agreement on the new "fact" was achieved by rhetorical means, and starts to think of the "fact" as "out there," i.e., something that really has all along existed in the natural world. However, this is merely self-deception, Latour and Woolgar contend. The "out there"-ness is just a figment of the scientific imagination induced by a sort of collective amnesia whereby scientists conveniently forget the real process of rhetorical manipulation that got everyone to accept the new "fact." I say "conveniently" because it is greatly in scientists' interests to portray themselves as discoverers of objective reality rather than, as Latour and Woolgar think they really are, just another set of tribal shamans pursuing their own myths and rituals. Biologist Matt Cartmill provides an unfriendly but accurate summary of the social constructivist view:

> The philosophy of social constructivism claims that the "nature" that scientists pretend to study is a fiction cooked up by the scientists themselves—that, as Bruno Latour puts it, natural objects are the *consequences* of scientific work rather than its *cause*. In this case, the ultimate purpose of scientists' thoughts and experiments is not to understand or control an imagined "nature," but to provide objective-sounding justifications for exerting power over other people. As social constructivists see it, science is an imposing but hollow Trojan horse that conceals some rather nasty storm troopers in its belly. (Cartmill, 1999, pp. 49–50; emphasis in original)

In his more recent writings Latour claims to have abandoned strict social constructivism. In *We Have Never Been Modern* (1992), he argues that scientific objects should be regarded as hybrid entities, neither as wholly natural, as scientists view them, nor as mere artifacts, as social constructivism holds. Rather, scientific objects, a virus, say, should be thought of as more or less natural or more or less constructed, depending on the context. Unfortunately, he never really clarifies just what it would mean to regard a virus as such a hybrid. Some commentators claim to have detected a "creeping realism" in Latour's later writings. They think that he begins to admit (sort of) that—just maybe—things like microbes really exist and have *some* bearing on the course of science. Whether or not this is so, with Latour, as with Kuhn, it was the earlier, more radical views that most impressed Latour's friends and critics.

Anyone imbued with a more traditional view of science might be tempted to dismiss social constructivism as a farrago of fuzzy thinking and exaggeration. The problem with such pat dismissal is that there are so many episodes in the history of science that do make us wonder how effective science really is at separating the real from the chimerical. How much of what we take to be facts about the natural world are really artifacts of our own making? One episode that raises this question is the famous wrongheaded dinosaur scandal.

The Carnegie Museum of Natural History in Pittsburgh houses one of the world's foremost exhibits of dinosaur fossils. For forty-five years the Carnegie Museum displayed one of its prize specimens, the gigantic *Apatosaurus louisae,* with the wrong head. The head was not a little bit wrong, but way off, like a paleontologist of the distant future putting a giraffe's head on a horse's body. Worse, other paleontologists accepted the chimera as real. All the top authorities accepted the wrongheaded creation as the real *Apatosaurus.* How did this happen?

Briefly, the problem had its roots in the feud between Marsh and Cope. Marsh published a reconstruction of *Apatosaurus* in 1883, which, to add to the confusion, he called by the familiar name *Brontosaurus,* mistakenly thinking it was a different kind of dinosaur than *Apatosaurus.* The problem was that *Apatosaurus/Brontosaurus* had been found without a head, and it just would not do to have a reconstruction without a head. So, Marsh improvised and stuck on a cranium that he had found at a completely different site. It eventually turned out that the head he found belonged to *Camarasaurus,* a creature not closely related to *Apatosaurus.* When the Carnegie Museum mounted its prize *Apatosaurus* specimen in 1915, the museum's director, W. J. Holland, an outstanding paleontologist in his own right, had deep reservations about the head Marsh had given the creature. Unfortunately, when the Carnegie Museum's specimen was found, it also lacked a head, so Holland had no definitive evidence against Marsh's reconstruction. Holland simply mounted the skeleton with no head.

Holland died in 1932 and in 1934 the new director of the museum decided that *Apatosaurus* needed a head. Just who made the decision and on what grounds is not clear. Probably, since museums are not only research institutions, but are there for the edification and entertainment of the public, everyone felt that a headless *Apatosaurus* just made a terrible impression. So, following Marsh's precedent, a very robust *Camarasaurus* skull was attached to the *Apatosaurus* skeleton. The big skull looked great on the massive skeleton and for over forty years everybody was satisfied, everybody except John S. McIntosh, perhaps the world's leading authority on sauropod dinosaurs like *Apatosaurus.* McIntosh began to suspect that another skull, one already in the Carnegie Museum's possession, was the right one. That other skull, designated as CM 11162, was found with the *Apatosaurus* specimen, but not in position at the end of its neck. Skull

CM 11162 looked too small to belong to so massive a creature as *Apatosaurus*. It looked like a somewhat larger version of a *Diplodocus* skull, and *Diplodocus* was a much slimmer animal than the ponderous *Apatosaurus*. However, after a thorough review of the records of the discovery of the Carnegie Museum's *Apatosaurus,* and a careful examination of skull CM 11162, McIntosh decided that it had to be the right one. He published his conclusions with coauthor David Berman in 1978. Their argument was so convincing that the Carnegie Museum agreed to remove the *Camarasaurus* head. Finally, in 1979, after forty-five years of displaying *Apatosaurus* with the wrong head, the Carnegie Museum held a ceremony to remove the old skull and attach a cast of CM 11162.

It is really quite shocking that one of the world's leading museums would display a prize specimen with the wrong head for so long. What if the Louvre had displayed a painting upside down for so long? Don't incidents like this make us wonder whether what we take to be scientific fact might not be artifact? Worse, there seemed to be no very good scientific reason for the Carnegie Museum's decision to mount the bogus head. It seems to have been a response to the demand to present a whole specimen for public viewing. Unlike the Venus di Milo, missing parts did not make *Apatosaurus* more appealing. Incidents like this lend credence to the claim that scientific decisions are often (strict constructivists would say *always*) made in response to social pressures, and not based on objective evidence.

Even if we admit, as we must, that science is often deeply influenced by social, political, and ideological pressures, must we accept the constructivist claim that, as Latour contends, the course of science is determined by power politics and rhetorical manipulation? Again, the issue of social and political influence on science is a very real and a very serious concern. It is indeed a serious matter when wealthy corporations aided by political ideologues manipulate or suppress legitimate scientific findings. But can the social constructivist theory be the *whole* story about science?

A problem with assessing the claim that facts are social constructs is that the word "fact" itself is ambiguous. According to *The American Heritage College Dictionary,* one sense of fact is "Information presented as objectively real." In another sense "fact" can mean "something having real, demonstrable existence." In other words, "fact" can refer to a *claim,* the assertion that something is really so, or it can refer to the *reality* that our factual claims are about. So, when one claims that scientific facts are constructed, this could mean either of two things: (a) our supposedly factual statements do not correspond to anything real (or if they do, it is a sheer accident), but are mere artifacts of the scientific process, or (b) there is no objective, "out there" reality consisting of states of affairs that exist independently of our beliefs or concepts. Sometimes Latour speaks as if he means to assert (a) and sometimes as if he means (b). So far, I have assumed that he means (a), since it seems a more plausible claim. That is, I have taken Latour to claim that, whether or not there is a real physical universe, such a putative natural world has no influence upon the "nature" conceived by scientists. The "nature" scientists study is therefore just an artifact, or, more bluntly, just a figment of the scientific imagination. Assuming that this is Latour and Woolgar's claim in *Laboratory Life*, how sound is it?

Latour and Woolgar present their conclusions as grounded upon empirical (i.e., scientific) evidence. Latour based his conclusions in *Laboratory Life* upon his research as an anthropologist of the laboratory. Anthropology is a science. If all scientific conclusions are social constructs, as *Laboratory Life* asserts, so are those of anthropologists like Latour. Practitioners of social constructivist anthropology or sociology of science have only two choices when they confront this problem of self-reference or "reflexivity" as it is called: They can bite the bullet and frankly admit that their "facts" are just as socially constructed as those of the natural sciences are alleged to be. In this case they have the burden of explaining why their so-called findings about the practice of science should be taken seriously by anyone skeptical of those alleged findings. The alternative is for social constructivists to argue, very implausibly, that social sciences like anthropology or sociology *do* draw upon reliable methods and objective evidence while the natural sciences, like organic chemistry or particle physics, do not. In other words, Latour's science was legitimate and Einstein's was not. Social constructivists have vigorously debated among themselves which horn of this dilemma to grasp, but neither option seems very appealing. The upshot is that if the social constructivist thesis is taken in the above sense (a), it is hard to see how constructivism can debunk science without debunking itself.

Sometimes, however, Latour seems to be making the above assertion (b)—that there is nothing that just *is* so, but that reality is, in some sense, created by the beliefs or concepts we form. That is, he sometimes seems to be implying a metaphysical claim about the nature of reality. For instance, in his book *Science in Action* (1987), he considers the famous case of the French scientist Rene-Prosper Blondlot, who, in the early twentieth century, claimed to have discovered a new type of radiation he called N-rays. The way the story is usually told, Blondlot thought he could observe a previously undetected sort of radiation emitted by metal under strain. Other physicists were skeptical because they could not reproduce Blondlot's claimed observations. An American physicist, Robert W. Wood, visited Blondlot's lab to see the procedure whereby N-rays supposedly were detected. At one point in his visit, while Blondlot was engaged with his experimental apparatus, Wood quietly removed an essential piece of the equipment. Yet Blondlot continued to proclaim that he could observe the N-rays as they were being generated. For Wood, and soon the whole physics community, this was proof that Blondlot's N-rays did not actually exist and that the reported "observations" of them were delusions.

Latour indignantly insists that we should not interpret this incident as implying that Wood was right and Blondlot wrong:

> It would be easy enough for scientists to say that Blondlot failed because there was "nothing really behind his N-rays" to support his claim. This way of analyzing the past . . . crowns the winners, calling them the best and the brightest and . . . says that the losers like Blondlot lost simply *because* they were wrong. . . . Nature herself discriminates between the bad guys and the good guys. But is it possible to use this as a reason why in Paris, in London, in the United States, people slowly turned N-rays into an arte-fact? Of course not, since at that time today's physics obviously could not be used as the touchstone, or more exactly since today's state is, in part, the *consequence* of settling many controversies such as the N-rays. (Latour, 1987, p. 100, emphasis in original)

That is, since the opinions of present physicists about the (non)reality of N-rays were shaped by the outcome of the N-ray controversy, those opinions cannot explain the outcome itself.

Well why not? Of course, it was, in part, Wood's fine job of debunking that convinced everyone at the time that Blondlot's claims were false. But once we are convinced that there was nothing there for Blondlot to detect, doesn't this explain why he failed to see them (given that his equipment *would* have detected them had they been there)? Why did people in April 1912 fail to see the *Titanic* docking in New York? Because it never arrived. As for why Blondlot *thought* he saw N-rays, this is given a psychological explanation in terms of how wishful thinking and the inherent limitations of human perceptual abilities can make us "see" things that are not there.

For Latour, such a common-sense account of the outcome of the N-ray episode just will not do. Notice his language. He says that people "turned" N-rays into artifacts. This seems to imply that for Latour, there was no fact of the matter, no way that things really were, before the controversy over N-rays was settled. It is not that Blondlot was deluded all along, and that Wood proved this to everyone else's satisfaction. Rather, Wood and others *turned* N-rays into artifacts. Does Latour think that N-rays could just as easily have been turned into real phenomena? It is hard to know just what Latour means here, but he seems to be saying that there simply was no fact of the matter about N-rays; they were neither real nor a mere artifact, until physicists *decided* the case. It is not just that no-body knew that Blondlot's purported observations were not real until Wood did his debunking; such a claim would be boring even if true (and Latour is never boring). Rather, there was NO fact of the matter—nothing there to know—until the physics community agreed on its story!

Now such a claim may strike many as bizarre, but it is not obviously in-coherent or self-defeating. It is reminiscent of the metaphysical idealism of British philosophy in the nineteenth century, which held that physical reality is a creation of the mind. I'm sure, though, that Latour would ab-jure any "idealist" label and certainly would disclaim any metaphysical agenda. Yet he seems to have fallen for an all-too-common fallacy of the sort that afflicted much idealist philosophy. This fallacious way of thinking begins with the innocent observation that we can only think with our ideas, but then leaps to the conclusion that all that we can know are our ideas. Michele Marsonet explains this fallacy and points out its obvious flaw:

> We do not know reality *directly,* but only through representations such as ideas and mental images. If this is true, it follows [so the fallacious argu-ment goes] that we only know our representations, while it is impossible to know an alleged reality in itself. However, it should be easy to realize that from the fact that we know *through* representations, it does not follow that we can only know representations and nothing else. . . . (Marsonet, 1995, p. 59)

Of course, the philosophical debate about the relation of ideas to reality is extremely long and complex, and we cannot enter it here even in the most superficial way. So, let us assume that Latour does hold, as he certainly seems to, that all that we can know are our own ideas, and cannot infer anything about any putative reality behind those ideas. He then has two options about how he thinks about any alleged mind-independent physical world: He can be agnostic about the existence of an objective physical universe, perhaps giving the stereotypical Gallic shrug when asked about it, or he can take the stronger position that there is no such world. We saw above that he sometimes speaks as though he takes the stronger line.

The problem with agnosticism about the existence of the physical world is that it is not clear how we are to explain the existence and the contents of our ideas unless we postulate physical objects as their causes. Surely René Descartes was right that our ideas have to be caused *somehow,* and do not just pop into existence *ex nihilo.* Immanuel Kant postulated the existence of unknowable "*Dinge an sich*" (things in themselves), as the cause of our perceptions. Later philosophers complained, reasonably enough, that we cannot meaningfully say that something exists—and much less that it causes all our perceptions—unless we attribute some sort of nature or character to it. So, it cannot be satisfactory to postulate physical objects as the unknowable "things in themselves" that cause our ideas. In *Laboratory Life,* physical objects do not even function as *Dinge an sich.* Instead, Latour and Woolgar invoke social and political factors as the sole and sufficient causes of scientific ideas. The problem with this option, as we have seen, is that it runs into the problem of reflexivity, that is, it undermines itself because if all truth claims are social constructs, so are those made by social constructivists.

Well, just what is wrong with denying the existence of the physical world? It is hard to say precisely how one could argue that there *is* an external, mind-independent physical world since any evidence you could mention would presuppose the existence of such a world. Could you convince someone like Latour of the reality of the external world by pointing to physical objects or waving them in his face? Should we attempt to prove the existence of an external world as British philosopher G. E. Moore famously did, by pointing to a hand and saying, "Here is a hand"? Steve Woolgar, Latour's collaborator on *Laboratory Life,* challenged his students to reveal a physical object to him without employing a representation of some sort. When students would point to a book or table, Woolgar would reply that pointing is itself a kind of representation. Maybe Woolgar was trying to make a point like the one artist Rene Magritte made when he painted a picture of a pipe and wrote under the depiction *"Ceci n'est pas une pipe"* ("This is not a pipe"). Just as the picture of a pipe is not a pipe, so pointing to a pipe is a representation, not a pipe. Woolgar was (I think) trying to show that we live in a world of representations, signs, symbols, images, and ideas—not objects.

What the students should have done is to point out that Woolgar has set them a task that is by definition impossible. You cannot indicate an object without indicating it, and Woolgar will say that any such act of indicating is a representation. But while it is certainly true that you cannot indicate something without indicating it, it does not follow that indicateable objects do not exist independently of our representations. Neither does it follow that we can never know objects, but only our representations.

The approach to take with questions like the existence of an external world is to begin by noting that, as the philosopher John Searle puts it, the human mind comes with certain default settings. When you first boot up a personal computer, in order to function at all it must come with certain default settings that it will keep until you change them. Likewise, the human mind seems just naturally pre-set to take certain things for granted. One of those things we just take for granted is that there is an external physical world that exists "out there" independently of our consciousness. Now default settings can be changed on your computer and in your mind, but we humans have every right to demand *very* good reason before we abandon a belief that is so spontaneous, natural, and (nearly) universal as belief in an external physical reality. In other words, to say the very least, a very heavy burden of proof is on those who would deny the existence of an external world. How could one make such an argument? Merely pointing out that we cannot think about things without using concepts, or indicate objects without somehow representing them, or make observations without appealing to some theory, will not prove this at all. All of these claims may well be true (even truisms), but it just does not follow from any of them that there is no mind-independent reality or that we cannot know a great deal about it. At this point I shall simply cut to the chase and assert that, in my opinion, nowhere does Latour, Woolgar, or any other social constructivist offer arguments anywhere near strong enough to support so sweeping a claim as the nonexistence of a mind-independent physical reality. Nor do they show that we fail to have cognitive access—*through* our perceptions and concepts—to that reality. So, Latour and Woolgar's version of social constructivism fails in its effort to debunk science.

POSTMODERNISM ATTACKS!

However, social constructivism is not the only recent program of radical science critique. There are also the postmodernists. "Postmodernism" is a term that defies precise definition. A variety of movements or styles in literature, art, and architecture may be called "postmodern." The "postmodernist" label has been attached to a wide variety of writers, including the philosopher Gilles Deleuze, his frequent collaborator the psychoanalyst Felix Guattari, sociologist Jean Baudrillard, psychoanalyst Jacques Lacan, and Luce Irigaray, whose writings deal with topics in many fields. So multifarious are these various manifestations of the postmodernist spirit that I can only give a very broad and impressionistic characterization of the attitudes and outlooks that tie them together. Anyway, postmodernist theorists themselves would probably reject any proffered canonical definition of "postmodernism," since one thing postmodernists share is a distaste for canonical statements of anything. For them, anything that presents itself as canonical, authoritative, or definitive is something to be abused, ridiculed, or otherwise subverted. For the postmodernists, it is a dangerous delusion to think that we ever have the complete story or final answer about anything. We are all ". . . on a darkling plain; swept with confused alarms of struggle and flight; where ignorant armies clash by night" (to quote Matthew Arnold, one of the canonical "great poets" postmodernists love to hate).

To avoid confusion, let me make an important distinction: Practically all philosophers these days are fallibilists. That is, they recognize that even our best-supported theories and factual claims are fallible and may turn out wrong—as, indeed, they so often have in the past. But fallibilism does not entail relativism. Even a thorough fallibilist can say that, so far as we can tell, some things just are so, some questions really have been settled, and some norms have at least *prima facie* validity. Not so for the postmodernists. For them all norms—whether ethical, aesthetic, or epistemological—have merely local authority and applicability and are radically contingent upon such factors as gender and social class. Postmodernists are hostile to any claim that a standard is more than merely parochial, viewing all claims to objectivity as attempts by one group to impose its values on others. Hence, in all fields postmodernists celebrate a promiscuous eclecticism of standards, values, and norms.

Postmodernism often comes across more of a style than a stance, more of a pose than a position. Sometimes postmodernists seem to take pride in outraging more traditional thinkers. They scorn even ordinary speech, which they regard as polluted by oppressive standards of clarity and truth. For them, the requirement that words should have definite meanings is just another tool of oppression. Their prose style therefore is often verbose, paradoxical, allusive, convoluted, and, in general, intended to disrupt and frustrate our ordinary ways of thinking. The downside is that much of what they say sounds like gibberish to outsiders. Postmodernists typically also reject the idea that it is a legitimate function of language to *represent* a language-independent reality. Baudrillard argues that an image should no longer be regarded as a simulation of reality, but that our media-drenched world *is* now a world of rootless, free-floating signs.

From what I've said so far, you may have gotten the impression that postmodernists are a bunch of zanies who lack seriousness of purpose and whose views have no intellectual motivation. Such an impression would be mistaken; postmodernists are entirely serious in their aims and their inspiration comes from thinkers who unquestionably were intellectual superstars. One such luminary was Friedrich Nietzsche (1844–1900). One aspect of Nietzsche's thought that the postmodernists have particularly emphasized is his view on the relation between knowledge and power. Nietzsche said that knowledge is an instrument of power, that is, that the motivation to acquire knowledge is to acquire more power. We want to "master" certain fields, thereby making that field of knowledge into a servant to promote our interests. For Nietzsche, as for the pre-Socratic philosopher Heraclitus, reality is an eternal flux that is always in a process of becoming. Knowing is not a matter of recognizing an objectively given reality; there is no such determinate, independent reality, only flux. Understanding therefore involves imposing our conceptual schemes, categories, and interpretations on the flux, thereby creating Being out of sheer Becoming. The interpretations we place on reality will reflect our vital interests and concerns, that is, the "reality" we create will be one that serves our purposes and enhances our power. Indeed, from an evolutionary perspective, only those "truths" that are useful survived. For Nietzsche, the idea that there is absolute truth, truth independent of all our interests and purposes, is a myth created by philosophers who hanker after a stable and permanent reality and are afraid to embrace endless flux. It follows that for Nietzsche, there is no one perspective, no "God's-eye view," that gives a comprehensive and authoritative view of the whole of reality.

Another, more recent, progenitor of postmodernism was philosopher Jean-Francois Lyotard (1924–1998). Lyotard argues that we should reject all "metanarratives." A "metanarrative" is any attempt to establish an absolute standard for any value or ideal—truth, rationality, goodness, or justice, for instance. Instead, we must recognize that there is an irreducible plurality of incommensurable narratives, each encompassing its own criteria for goodness, truth, etc. Lyotard expressed these views most influentially in his work *The Postmodern Condition: A Report on Knowledge* (1979).

Another notable philosophical work published in 1979, Richard Rorty's *Philosophy and the Mirror of Nature* articulated and promoted various postmodernist themes for an English-speaking audience. Rorty characterizes his philosophy as "pragmatist," i.e., in the tradition of classical American philosophers such as William James and John Dewey, but many of his ideas are typically postmodernist. For instance, he strongly advocates that philosophers abandon the attempt to establish absolute foundations for knowledge and instead dedicate themselves to the facilitation of the "conversation of mankind." According to Rorty, all of the "voices" of humanity, from Hopi philosophy to Polynesian mythology to quantum physics, deserve to be heard and no one narrative should be preeminent. When it comes to grounding our beliefs, Rorty says that we can do no better than to say what our society lets us say, that is, when in Rome, follow the epistemological practices of the Romans. This does not mean that our beliefs are not to be subject to strict critical scrutiny, but Rorty thinks that the standards we employ when we thus examine our beliefs are contingent historical products of a particular time and place and lack universal authority.

It is hardly surprising, given such an intellectual background, that postmodernists do not like science very much. After all, there are some things that science says are just so, and others definitely not so—period. Heavy bodies in free fall in the vicinity of the earth's surface accelerate at a rate of about 9.8 meters per second squared. The sun has a mean distance from the earth of 149.5 million kilometers. The nuclei of human somatic cells contain 46 chromosomes arranged in 23 pairs. Science does not say that such things are so from a given perspective, or according to some traditions, but that they are just so. Science claims to be the authority in answering certain questions. Where did birds come from? They evolved from theropod dinosaurs in the late Jurassic, say (many) paleontologists. Science says that the story that God created them all at once during the six-day creation is simply false. Further, science claims that its methods alone are the right methods for investigating the natural world and not, for instance, consulting horoscopes, gazing into crystal balls, or invoking the authority of ancient texts.

Because science so often claims to have *the* answer, and not simply to offer one among indefinitely many perspectives, it is bound to ruffle postmodernists' feathers. What right do paleontologists have to tell Christian fundamentalists and Orthodox Jews that the Genesis creation account is false? Is it not arrogant for anthropologists to tell Native Americans, whose traditions teach that they are indigenous to North America, that their ancestors actually came across a land bridge from Asia? Following Nietzsche, postmodernists say that knowledge is power. They do not mean this the way that Francis Bacon did, as a recognition of the fact that knowledge gives us power over nature, but in the sense of one of their favorite writers Michel Foucault (1926–1984). When Foucault says that knowledge is power, he means it in the sense that the winners get to write the history books. For instance, the anthropological account of the origin of American Indians is just the story that the winning white European culture gets to impose on the losing Native American culture. Postmodernists regard the reigning standards that define rational discourse—the standards that tell us what counts as a logical inference, objective evidence, or coherent speech—as potential tools of oppression. Small wonder that postmodernists want to challenge what they see as the intellectual hegemony of science. Though their jargon may be opaque, their intentions are clear. They aim to cut science down to size, to display it as just another form of discourse, and as no more "rational" or "objective" than any other. Postmodernist literature is vast and highly diverse. From these many writings I have selected two books to examine here: Donna Haraway's *Primate Visions* (1989) and W. J. T. Mitchell's *The Last Dinosaur Book* (1998). These two books offer postmodernist analyses of two fields of science that have much popular appeal, primatology and dinosaur paleontology.

Reading postmodernist literature, you can get the impression that they are obsessed with the electronic media. Because they reject the distinction between "high" and "low" culture, postmodernists repudiate the traditional academic disdain for popular entertainment. Papers by academic postmodernists will often have titles like *From Homer to Homer Simpson* where canonical texts like the *Iliad* and popular TV comedy get equal (and equally obscure) treatment. Not only do they collapse the distinction between popular and highbrow entertainment, they go even deeper, questioning the very distinction between a symbolic representation and the reality that it represents. Typically, postmodernists oppose what Haraway calls "binarisms"—paired concepts that, in their view, channel our thinking into narrow and misleading dichotomies. They say that rigid distinctions like subject/object, fact/fiction, same/other, and image/reality are embedded in our language, and so lead us to box things into overly restrictive categories. Invidious value judgments go with such labeling, postmodernists argue, so that, for instance, people the "same" as us are good and those "other" than us are not.

Haraway attempts to subvert the distinction between science fact and science fiction. Likewise, Mitchell argues that it is impossible to maintain the distinction between the popular image of dinosaurs and what paleontologists think they really know. For Haraway, science is just another kind of narrative. True, primatologists have their story to tell, but it is just one of many and has no special authority over any other account. Similarly, Mitchell certainly feels that paleontologists should lend their "testimony" to our understanding of dinosaurs, but such "testimony" must be supplemented by the work of humanities scholars who are experts at the analysis of symbols and images. After all, echoing some of Latour's talk about hybrid objects, Mitchell says the dinosaur is not merely a natural object nor is it a pure fantasy, but is an irreducible composite. Mitchell holds that there is no way to make a workable demarcation between the extinct animal and the cultural icon.

For Haraway, the idea that science can be done in a neutral, disinterested, and impartial way is a pernicious myth. It is a myth because, she holds, all science is inevitably *political* science, since it always promotes the interests of some particular group. The myth of neutrality is pernicious because it obscures the fact that every scientific account, however purportedly "objective," or based on "logic," serves a hidden agenda. So far that agenda has been the promotion of the interests of scientists, and their sponsors in government and industry, who are almost always white, male, and privileged. She strongly endorses the Latour/Woolgar view that all scientific "facts" are socially constructed and adds that it is vital to see for whose benefit they are constructed. She cites a well-known story about the history of primatology. According to this story, when primatologists began to analyze the social structure of primate groups, the (predominately male) scientists focused on the dominance of the so-called alpha male. The alpha male, like the silverback leader of a gorilla troop, is the dominant male. According to those early accounts, the story goes, the dominance of the alpha male was depicted as absolute. In particular, all of the females of the group were, in effect, the harem of the dominant male since he had exclusive right to mate with them.

It is easy to see how such a representation of primate social groups could benefit males. Since primates are the closest animal relations to humans, the alleged dominance of the alpha male could be seen as the natural pattern for human society as well. That is, male dominance in human society could be justified as "natural" by pointing to the dominance of the alpha male in primate society. However, says Haraway, it fell to female primatologists to point out that the alpha male's dominance is far from absolute, and that female primates wield considerable power. For instance, among mandrills when a new alpha male takes over by defeating the previously dominant male, he finds, no doubt to his intense chagrin, that the females are not instantly his to command. The females will defiantly refuse to mate with him until he meets their approval. Of course, Haraway cannot think that the female primatologists' conclusions were any more objective than those of their male counterparts. What matters about the stories science tells is not whether they are "true" (the quotes are needed because postmodernists do not buy the notion of just plain truth). What matters is whose interests those stories serve. As we shall see below, feminist theorist Sandra Harding picks up on this theme and runs with it.

Science has always assumed that the objects it studies are determinate entities that exist objectively and independently of the merely human activity called science. Science therefore gave itself the job of discovering such entities and understanding them as fully as possible. Even the weirdness of quantum mechanics has not really altered this fundamental goal of science. Quantum mechanics, at least as it is usually interpreted, tells us that there are some properties of particles that have no definite values until we interact with those particles in some way. But once the requisite interaction occurs and the particle assumes a definite value, then that value is as objective and determinate a fact as any other. For postmodernists, scientific objects lose all such status as determinate and independent entities. In postmodernist literature, a scientific object is little more than a nexus of multiplying interpretations, a blank screen onto which interested parties may project practically any image.

For paleontologists, a dinosaur was an *animal,* a creature that roamed the Mesozoic landscape and possessed distinct anatomical, physiological, and behavioral traits that we try to discover by the framing and testing of hypotheses. Mitchell treats dinosaurs as prefabricated metaphors, ready-made symbols that can stand for just about anything. For instance, dinosaurs can symbolize obsolescence, backwardness, and stupidity. Cartoonist Gary Larson picked up on this theme in his *The Far Side* strip. He depicted a Stegosaur lecturer speaking to an audience of dinosaurs: "The picture's pretty bleak, gentlemen. . . . The world's climates are changing, the mammals are taking over, and we all have a brain about the size of a walnut." On the other hand, dinosaurs can be cool and chic, like the sleek, fast, and deadly *Velociraptor* of *Jurassic Park*. According to Mitchell, dinosaurs can serve as the "clan sign" for just about any group. *T. rex* could symbolize unbridled ferocity while *Apatosaurus* might represent the gentle giant. A dinosaur can even be a plush purple TV figure who warbles saccharine ditties to preschoolers. For Mitchell, any attempt to strip away the layers of symbolism and get down to the *real* dinosaur would be like peeling an onion. You would never hit factual bedrock, only layer upon layer of symbol and metaphor. The upshot is that paleontologists cannot hope to understand dinosaurs, since they are under the illusion that they are studying an unambiguously *natural* object. Mitchell argues that to really understand dinosaurs, the researches of paleontologists must be supplemented by the work of humanities scholars, like himself, who are experts at the analysis and interpretation of symbols. Since dinosaurs are hybrid objects, irreducible composites of the natural and the symbolic, they must be studied by a hybrid discipline that combines the "testimony" of paleontologists with the interpretations of practitioners of cultural studies.

For Mitchell as for Haraway, scientific objects *always* have political overtones. For instance, he sees the paintings of battling dinosaurs done by Charles R. Knight at the beginning of the twentieth century as symbolic of the unrestrained capitalism of the Gilded Age:

> Knight's scenes of single combat between highly armored leviathans are the paleontological equivalent of that other war of giants, the struggles among the "robber barons" in late Nineteenth-Century America. This period, so often called the era of "Social Darwinism," economic "survival of the fittest," ruthless competition and the formation of giant corporate entities headed by gigantic individuals, is aptly summarized by the Darwinian icon of giant reptiles in a fight to the death. (Mitchell, 1998, p. 143)

Postmodernists also often claim to detect a sexual subtext in contexts where to others it seems hardly present. For instance, he comments on Henry Fairfield Osborn's dinosaur displays at the American Museum of Natural History in the early 1900s:

> Perhaps Osborn's most important contribution to the myth of the modern
> dinosaur was his linkage of it to questions of male potency. The connec-
> tion between big bones and virility had already been established. . . . Big
> bones were also the trophies of the masculine ritual of the big game hunt,
> and the phallic overtones of "bones" need no belaboring by me. (p. 150)

Even the greenish color given to dinosaurs in most depictions is full of
symbolic import for Mitchell:

> So where does this leave greenness? Is it a symbol of the "colored" racial
> other, the savage, primitive denizen of the green world? Or is it an em-
> blem of the white man's burden, the color of the military camouflage re-
> quired for the Great White Hunter to blend in with the jungle and thus to
> dominate it? (pp. 147–149)

Can *everything* about dinosaurs really be bursting with political and/or
sexual significance? Mitchell apparently thinks so, and he therefore recommends that Marx and Freud be invoked to analyze the political and
sexual content of dinosaur images.

Any attempt at a straightforward point-by-point rebuttal of postmodern
critiques of science will probably fail. This is not because those critiques
are sound and therefore irrefutable. Rather, it is because almost anything
a critic would take as a flaw of postmodernist analyses would be seen as a
virtue by the postmodernists. Leading primatologist Matt Cartmill vents
his frustration in attempting to criticize Haraway's *Primate Visions:*

> This is a book that contradicts itself a hundred times; but this is not a crit-
> icism of it because its author thinks contradictions are a sign of intellec-
> tual ferment and vitality. This is a book that systematically distorts and
> selects historical evidence; but that is not a criticism, because its author
> thinks that all interpretations are biased, and she regards it as her duty to
> pick and choose her facts to favor her own brand of politics. . . . This is a
> book that clatters around in a dark closet of irrelevancies for 450 pages
> until it bumps accidentally into an index and stops; but that's not a criti-
> cism, either, because its author finds it gratifying and refreshing to bang
> unrelated facts together as a rebuke to stuffy minds. . . . In short, this book
> is flawless, because all its deficiencies are deliberate products of art. (Cart-
> mill, 2003, p. 196)

Perhaps we have at last found a genuine example of incommensurable
discourse: the debate between postmodernists and their critics!

Seriously, though, how do you meaningfully disagree with those who
have apparently repudiated the very conditions of meaningful disagreement? How do you deploy objective evidence against those who regard objectivity as a myth? Perhaps the would-be critic would try to "deconstruct"
postmodernist texts. "Deconstruction" is a kind of radically skeptical

textual analysis frequently used by postmodernists. A deconstructive analysis turns a text against itself and attempts to show that it has no definite, distinct meaning, but lends itself to innumerable interpretations. Could we deconstruct Haraway's and Mitchell's texts? There seems to be no reason why not. That is, if we had the patience, we could no doubt go through their texts and pick out numerous passages that we could then interpret as meaning the exact opposite of what Haraway and Mitchell apparently intend. For instance, we could take Haraway's animadversions against objectivity and interpret them as ironical *defenses* of objectivity that work by showing the absurd consequences that follow when objectivity is repudiated. Likewise, we could take Mitchell's meditation on dinosaurs' greenness as a demonstration of the silliness that inevitably results when basic distinctions are systematically conflated, like the distinction between an object and its image. In short, it looks like postmodernists are vulnerable to the same problems of reflexivity that plagued the social constructivists. If all texts can be deconstructed, so can the texts of postmodernists.

But such a quick, down-and-dirty dismissal of postmodernism completely misses the point since postmodernists emphatically reject the canons of rationality that underlie any such critique. They reject all demands that texts meet standards of consistency, coherence, or truthfulness. Postmodernists have no problem with reflexivity. They would be the first to admit that their own texts can be deconstructed! Perhaps then postmodernist texts should not be regarded as rational arguments; their goal is not to arrive at truth, or even to achieve coherence; such notions are for them passé. Postmodernist writing is above all a *performance*. That is, perhaps it is best to regard postmodernist science critique as a genre of confrontational performance art; its goal is not to persuade but to provoke. Some critics of postmodernism have therefore concluded that instead of wasting rational argument on such provocateurs they should play tricks back on them.

This is precisely what physicist Alan Sokal did when he wrote a spoof of postmodernist science critique, intentionally filled it with arcane postmodernist jargon and absurd arguments, and passed it off as a serious article to *Social Text*, a periodical that prominently features postmodernist writers. Sokal gave his piece a suitably portentous title: "Transgressing the Boundaries: Towards a Transformative Hermeneutics of Quantum Gravity." The text was a farrago of ludicrous claims about the political implications of recent developments in physics spiced with particularly opaque passages from leading postmodernist writers. In 1996 *Social Text* published Sokal's parody as a serious article, and the joke was on them. Sokal revealed the hoax in the periodical *Lingua Franca,* and contended that his successful sting had exposed the ignorance and laziness of the postmodernist science critics. He charged that such critics had shown that they would endorse anything, no matter how incompetent, that supported their view. Needless to say, many of the postmodernists were embarrassed and outraged and responded to Sokal with considerable asperity. Stanley Fish, a noted literary scholar and onetime editor of *Social Text,* castigated Sokal and accused him of creating a spiteful Trojan Horse to embarrass colleagues. Such behavior, Fish charged, only undermined the basic trust necessary for scholarship as a cooperative enterprise. Even some philosophers of science who are sympathetic to Sokal's view feared that his hoax would only lead to polarization when bridge-building between various disciplines is sorely needed.

There are places in postmodernist writings where they do seem to be making straightforward claims backed by evidence and argument. For instance, what are we to make of Mitchell's proposal that paleontology be replaced by a hybrid discipline that combines the expertise of paleontologists and the skills of "cultural scientists," as he thinks specialists in his field should be designated? Our reaction to this proposal will depend on how we take Mitchell's claim that dinosaurs are inevitably hybrid objects and that it is impossible to scrape off the accretion of symbolism and get down to rock-solid, literal truth about dinosaurs. Now admittedly there are some things about dinosaurs we do not and very probably never will know. For instance, the colors of dinosaurs will probably remain conjectural. We just have no way of knowing whether dinosaurs were the greenish color that got Mitchell's interpretive juices flowing or whether, maybe, they really were purple. We hardly know everything about living creatures, so how could we ever know everything about extinct ones?

Yet we seem to know that some things about dinosaurs *are* so, and Mitchell never offers any good reason to doubt that we do. Just because an object has potent symbolic import for us is no reason to think that we cannot know many things that are literally true of that object. For instance, a cross naturally has deep symbolic significance for devout Christians, but Christians can still understand the cross as the instrument of torture and death that it actually was. Mitchell tells us "Nature *is* culture, science is art. We don't ever 'see nature' in the raw, but always cooked in categories and clothed in the garments of language and representation" (p. 58; emphasis in original). Of course, since it is true by definition, we must admit that we cannot think about nature without using language, categories, and representations (we cannot think about something without thinking about it). But we can admit this and still think that we do, on occasion, get things right.

How would the workaday scientist react to postmodernist writers? He or she would probably think that writers like Haraway and Mitchell, whose academic careers have involved them exclusively in a world of symbols, tropes, and texts, have simply lost contact with the intractable, obstinate, downright recalcitrant world of physical fact that scientists confront daily. Even scientists who never leave their air-conditioned labs must struggle daily to square their conjectures with the hard constraints imposed by an unyielding cosmos. When it comes to telling stories about dinosaurs, Mitchell says, "There is no limit to the stories that can be made up. . . ." (p. 48). For the paleontologist, coming up with even *one* story can be devilishly difficult. The reason for this difference is that nothing constrains Mitchell's storytelling except the limits of his own imagination. Paleontologists' stories are severely constrained both by background knowledge and by physical fact. For Mitchell, the stories we tell about dinosaurs should be full of "fantasy, unbridled speculation, and utopian imagination" (p. 284). Science also thrives on speculation and imagination, but in science fancy must sometimes be allowed to crash into the hard rock of empirical reality. Because of these fundamental differences between Mitchell's cultural studies approach to dinosaurs and the paleontologist's, a hybrid discipline that yokes these two disciplines is not feasible.

IS "OBJECTIVITY" WHAT A MAN CALLS HIS SUBJECTIVITY?

A major intellectual movement of the last three decades has been the rise of feminist scholarship. Science has drawn the attention of many feminist writers. These writers certainly found much about science that rightly concerned them. When you look at the index of any history of science, you find that scientists have been a very diverse lot. Over the past 5,000 years significant scientific discoveries have been made by Egyptians, Babylonians, Greeks, Chinese, Indians, Arabs, Jews, Mayans, Italians, Germans, English, Scots, Russians, Hungarians, French, Danes, Americans . . . and on and on. Great scientific work has been done by Pagans, Christians, Jews, Muslims, Hindus, Buddhists, Confucians, Atheists, and Agnostics. Scientists of my personal acquaintance run the gamut from conservative Republicans to Marxists. Scientific journals are published in dozens of languages. In a given year, a scientist might attend professional conferences in London, Rio de Janeiro, or Riyadh. A *Who's Who* of scientists would have representatives of almost every race or ethnicity. Truly, science seems to be a characteristically human enterprise and not the domain of just one group.

Or is it? One jarring fact about the names we find in an index of the history of science is that the overwhelming majority of those scientists were *male*. Of course, there have been distinguished women scientists from ancient times to the present day. The list would include such names as Hypatia, Caroline Herschel, Marie Curie (winner of Nobel Prizes in chemistry *and* physics), Irène Joliot-Curie, Lise Meitner, Ceceliu Payne-Gaposchkin, Barbara McClintock, Rosalind Franklin, Vera Rubin, and Lynn Margulis. But these were very much the exception to the rule. One reason for the lack of female names in the lists of notable past scientists is unquestionably that women in the past simply were not given due credit for their scientific work. For instance, Caroline Herschel was certainly a notable astronomer, but her work is usually mentioned as a footnote to the achievements of her more famous brother William Herschel. Still, there can be no doubt that the apparent paucity of women scientists largely reflects the historical reality of women's exclusion from science. The inescapable conclusion is that over the centuries a vast reservoir of scientific talent was hardly tapped at all. Things are better today, but still far from ideal. Some fields, like medicine, have basically achieved parity in numbers (if not in status and power); other areas, such as many engineering fields, are still over 80 percent male.

Why has science historically been, and to a large extent still is today, so predominately a male enterprise? For feminists, the answer is obvious: Science is a boy's club. It is run not only by but for men. Women are still subtly, and sometimes not so subtly, discouraged from entering science. Even when women succeed in becoming scientists, they sometimes are marginalized and relegated to lower-status jobs or just "left out of the loop" by male colleagues. A woman of my personal acquaintance, a nuclear engineer, was told by her boss that women are "weak links" who must be driven from the profession. Further, male scientists have certainly entertained some odd, even bizarre, notions that it is hard to imagine female colleagues—had there been any—taking seriously. For instance, many physicians of the nineteenth century addressed the problem of hysteria, allegedly a nearly universal ailment of women that was supposed to cause them to experience uncontrollable emotions (*hysteria* is the Greek word for "womb"). The diagnosis of a troublesome woman's behavior as hysterical was certainly a convenience for men. If a woman made a public scene (like demonstrating for the right to vote), the problem was medical; she was hysterical.

Feminists are rightly concerned about such issues. Insofar as they aim to redress past wrongs and assure that women have equal opportunities to enter and advance in scientific fields, their efforts can only be welcomed. But for more radical feminists, such efforts merely scratch the surface. The deeper problem, as they see it, lies not just with the obstacles that have been put in the way of girls and women who might enter science, but with the very ideals and standards of science. The focus of much radical feminist science critique is the idea, which scientists have championed since the scientific revolution of the seventeenth century, that science should be value free. Here we must make an important distinction between epistemic values and non-epistemic values. As we noted in an earlier chapter, "epistemic" means "relating to knowledge," so, the epistemic values of science are those conducive to the aim that science produce genuine knowledge. For instance, scientists place great value on the rigorous empirical evaluation of hypotheses since they hold that only such stringent testing can eliminate false hypotheses and lead us towards the true ones. Non-epistemic values are those which, although most important for human life, do not set norms for good inference or the correct evaluation of evidence. For instance, moral values can tell us that murder is wrong, but they do not tell us the right way to conduct a homicide investigation. A homicide investigation is an empirical inquiry and is guided by epistemic values. Non-epistemic values also include political values, spiritual values, aesthetic values, and so forth.

Traditionally, scientists and philosophers of science have held that science should, insofar as possible, make its practice independent of non-epistemic values. The reason why seemed obvious. People are passionate about their moral, political, and religious values. The "hot button" issues that enliven political campaigns and editorial pages are hot because they involve such basic values. Budget deficits might doom our children to a future life of hardship, but that distant prospect gets people less excited than whether kids should say "under God" while reciting the Pledge of Allegiance. Scientists are people too and just as likely as anyone else to have strong feelings about political, religious, and moral issues. If, therefore, the results of science were not made independent of our strong feelings about moral, political, and religious values, scientists feared that scientific objectivity would be badly compromised. Objectivity is the goal of telling it like it is, even when our scientific conclusions run against deeply entrenched convictions.

Unquestionably, science does often conflict with such entrenched convictions. Darwinism is probably the most obvious example. There is the famous anecdote about the aristocratic Victorian lady's reaction upon first hearing of Darwin's theory. She cried out to her husband in horror: "Oh, my dear! Descended from apes! Let us hope that it is not true, and if it is that it does not become generally known!" Philosopher Daniel Dennett has rightly characterized Darwinism as "universal acid," an idea so corrosive that it threatens to dissolve any dogma or ideology it contacts. The only problem with Dennett's claim is that it is not broad enough. Any scientific theory, not just Darwinism, can and often does undermine entrenched beliefs. Inevitably, then, science will often run into entrenched ideological opposition and the inevitable pressure to modify or reject scientific conclusions when it does. To prevent such ideological obstruction, and to permit science to tell it like it is even when the truth hurts, scientists have embraced norms and methods intended to identify the polluting taint of ideology and insulate science from its influence. For instance, though scientists like everyone else often have strong political convictions, they are supposed to follow norms and practice methods that prevent their convictions from distorting their science. Thus, science is supposed to be dispassionate, disinterested, and politically neutral even though the people that practice it, being merely human, cannot be.

Many feminist writers reject the idea that science should, or can, be freed from the influence of non-epistemic values. They argue that science is inevitably and pervasively influenced by such values and that these shape both the practice and the conclusions of science. It is therefore pointless, and in fact disingenuous, for science to pretend to be value free (from now on, by "value," I'll mean non-epistemic value). Rather, it should explicitly embrace the *right* values. Science should begin to serve the interests of the down-and-out, and stop catering to the up-and-in.

Sandra Harding is one feminist writer who argues this way, and she gives this argument perhaps its most uncompromising expression. She says that science has so far claimed to follow the ideal of neutral, disinterested, and dispassionate inquiry. She calls this ideal "weak objectivity" and says that, though scientists and philosophers of science pay lip service to this ideal, in reality it never has and never will be actually practiced. She cites Kuhn and the social study of science to back her claim and concludes:

> Modern science has again and again been reconstructed by a set of interests and values—distinctively Western, bourgeois, and patriarchal. . . . Political and social interests are not "add-ons" to an otherwise transcendental science that is inherently indifferent to human society; scientific beliefs, practices, institutions, histories, and problematics are constituted in and through contemporary political and social projects and always have been. (Harding, 1991; 2003, p. 119)

In short, Harding endorses the slogan that heads this section: "Objectivity [as traditionally construed] is what a man calls his subjectivity." That is, the values that guide science have been those that men personally endorsed because they served male interests. Men tried to disguise the self-serving and subjective nature of those values by calling them "objective," "disinterested," and "impartial." Further, the influence of social and political factors on science cannot be blocked by adopting stricter methods and tighter controls. The idea of a value-free science is therefore not only a myth, it is a dangerous myth since it employs the language of impartiality to obscure the hidden agendas that science has always served.

Harding recommends that science instead pursue the ideal of "strong objectivity," which science can achieve only when it starts "thinking from women's lives." That is, "women's experience," specifically as interpreted by feminist analysis, must inform all of the standards, values, and methods of science. Feminist analysis turns women's experience of sexist oppression into a source of insight by raising victims' consciousness. Men, as the beneficiaries rather than the victims of sexist oppression, will not have such insights. It is like the situation where the slave knows all of the master's moods, whims, and quirks perfectly, but the master is largely oblivious of his slaves' lives. To get the benefit of women's experience (correctly interpreted), and realize the ideal of "strong objectivity," science must explicitly adopt the feminist standpoint.

Let me emphasize here that Harding is not making the rather bland recommendation that science should be open to people from many different backgrounds so that scientific communities will contain persons with a variety of perspectives arising from differences of "life experience." The idea that there should be a diverse "web of knowers" whose different perspectives will lead to a fairer evaluation of knowledge claims seems to be a reasonable suggestion. Harding, though, wants science to adopt a *specific* perspective and set of values—hers.

Now Harding realizes that critics will point to notorious cases when people adopted doctrinaire assumptions and tried to do "politically correct" science. Perhaps the most famous incident involved Soviet pseudoscientist Trofim Lysenko, whose attempt to introduce a Marxist/Leninist brand of genetics destroyed the legitimate practice of that science in the Soviet Union. By currying favor with Stalin, who imposed a totalitarian stranglehold on every aspect of Soviet life, Lysenko was able to get his rivals banished to the Gulag and insure that only his own views were taught. The attempt to apply Lysenko's crackpot theories to agriculture resulted in disaster, and, perhaps, ultimately contributed to the fall of the Soviet Union. Maybe equally notorious were the attempts in Nazi Germany to pursue an "Aryan Science" that repudiated "Jewish Science" such as the theories of Einstein.

Harding's reply is to reiterate her allegation that a value-free, apolitical, impartial science is a myth, and again to state that all science must be done from some political perspective and work to promote some set of values. Therefore, we must choose whether science will serve a liberating set of values or the values of an oppressive doctrine. Immediately, though, there is a problem. Just who gets to decide which doctrines are "liberating" and which are not? Why cannot evangelical Christians, for instance, insist that *their* doctrine is the liberating one and that science should be based on *their* perspective and values? Harding says that the feminist standpoint is preferable because it leads to a science that is less "partial and distorting" than other doctrines. However, this implies that we can have some impartial means of telling which doctrines are more "partial and distorting." That is, we have to have some dependable methods for recognizing whole, undistorted truth and distinguishing it from the half-truths and distortions of ideologues. Those methods, whatever they are, of distinguishing truth from distortion cannot themselves presume the feminist standpoint. Assuming the feminist standpoint to prove the legitimacy of the feminist standpoint is obviously arguing in a circle. If Harding admits this, she must admit that we have reliable means of recognizing whole, undistorted truths, and that those means do not depend upon adopting the feminist standpoint or any other ideology. But this seems tantamount to admitting that we *can* have a disinterested, impartial, value-neutral science.

A deeper issue is what exactly it would mean for science to adopt the feminist standpoint. Most important scientific discoveries were credited to male scientists. It is a good bet that most of these men of science shared the common prejudices of their day. Some were even blatantly misogynistic. However, with many of these discoveries it is very hard to see how the incorporation of "women's experience" into the information available to scientists would have aided the discovery process or made the results less "partial and distorting." Objects fall at the same rate for women as for men. A cup of hot coffee cools at the same rate for women as for men. The speed of light is the same for both men and women. It is hard to think that Newton's laws and principle of universal gravitation, the laws of thermodynamics, or the principles of special relativity would have been any different had they been formulated by feminists. Perhaps Harding would concede this, and she would now say that it is only in the more "human" and social sciences such as anthropology, sociology, primatology, psychology, and so forth that adopting the feminist standpoint would make a significant difference.

Now it is certainly true that bias of various sorts has at times distorted science. Harding is quite right that the banner of scientific objectivity has often been unfurled to hide the ugliness of bigotry. Stephen Jay Gould's 1981 book *The Mismeasure of Man* is both amusing and horrifying when it recounts how nineteenth-century anthropologists pursued craniometry, the measurement of the size of human skulls. They simply assumed that a bigger skull would house a bigger brain and therefore indicate higher intelligence. They measured the cranial capacity of the skulls of many races—Europeans, Sub-Saharan Africans, Eskimos, Semites, Native Americans,

Australian Aborigines, and East Asians. Which group did these European researchers find to have the greatest cranial capacity and therefore the highest intellectual ability? You get one guess. German researchers even found that German skulls were more capacious than other European skulls. Other prejudices have distorted other sciences. Until the 1970s homosexuality was listed in psychiatry texts as a mental disorder. The silly things that male scientists have said about women could (and did) fill volumes. So, would these sciences have been made more objective had they adopted the feminist standpoint?

The answer depends, first, on whether Harding has shown that all science is inevitably and inextricably bound to political agendas and the promotion of non-epistemic values. If she has not, then perhaps the way to remove bias from science is not to bring in a new ideology, but to pursue the old-fashioned goal of a more impartial science. Second, we have to ask whether the feminist standpoint would itself introduce its own form of bias so that there will be no net gain in achieving a less "partial and distorted" science. Quite frankly, there are tenets of feminist doctrine that could bias science by placing an *a priori* ban on possible scientific results. For instance, perhaps the leading school of feminist thought today is called "gender feminism." Gender feminists argue that, while sex is a biological fact, gender is a social construct. That is, it is a natural fact that women bear children and men do not, but the various social roles that have traditionally been assigned to women and to men are, in their view, entirely products of culture. As gender feminists see it, the stereotypes about men and women's behavioral dispositions, such as that men, as a matter of biological fact, tend to be more physically aggressive and women more nurturing, are all false. In their view, in a restructured society, one engineered around feminist values, these supposedly "natural" behavioral differences would wash out.

But what if gender feminists are wrong on these points? What if there really are natural differences in the behavioral dispositions of the sexes? Of course, some women are more physically aggressive than some men, just as some women are taller than some men. But what if, just as men are, on average, naturally taller than women, they also, on average, are naturally more physically aggressive? Steven Pinker in his recent book *The Blank Slate* (2002) argues very cogently that there are innate, biological dispositional differences between the sexes. Pinker could be wrong; debates over these points are far from settled. But the point is that he could be right. There is no way we can know ahead of time; we have to carry out the research and see. If Pinker is right, that is if the natural facts are as he says, then gender feminists might be casting themselves in the role played by Pope Urban VIII when he proscribed Galileo's findings because they contradicted sacrosanct beliefs. Even if gender feminism turns out to be consistent with all empirical findings to date, we should continue to presume it true only so long as it continues to face all challenges. John Stuart Mill (himself a strong advocate of feminism) spoke what should have been the final words on this matter:

> There is the greatest difference between presuming an opinion to be true, because, with every opportunity for contesting it, it has not been refuted, and assuming its truth for the purpose of not permitting its refutation. Complete liberty of contradicting and disproving our opinion is the very condition which justifies us in assuming its truth for purposes of action; and on no other terms can a being with human faculties have any rational assurance of being right. (Mill, 1952, p. 276)

It is hard to avoid the impression that Harding would be most displeased if science had "complete liberty of contradicting and disproving" her version of feminism.

IS SCIENCE GODLESS?

The science critics we have so far considered—social constructivists, postmodernists, and gender feminists—are representatives of the "academic left." Postmodernism, for instance, seems to have sprung from the failure of the French radical movement of the late 1960s. Yet it should come as no surprise that conservatives have also recently challenged the claimed objectivity of science, at least as it is now practiced. Science has traditionally clashed more often with conservative thinkers than left-wing ones. The most notorious conflicts occurred when science and religion clashed. Science and religion are not always opposed; sometimes they even cooperate in a symbiotic relationship. Yet conflicts are inevitable. Stephen Jay Gould, the author of *The Mismeasure of Man* mentioned above, argues that science and religion do not clash because they are what he calls "non-overlapping magisteria." That is, science deals with matters of physical fact and theory while religion is the realm of value and spirit. But this is merely wishful thinking. There is no self-evident principle that relegates science and religion to different spheres. There is no *a priori* reason why religion cannot have something to say about the physical universe or why science cannot say something about value.

In fact, there are any number of ways that science and religion can clash. For instance, many religions teach that humans have an immortal soul, the seat of mind and consciousness, that will survive (and, according to some traditions, predates) its incarnation in a human body. Yet the flourishing field of neuroscience takes it for granted that mind and consciousness, those phenomena previously thought to be the province of the soul, are due entirely to the physical functions of neurons—brain cells. Of course, neuroscience has not, and may never, explain precisely how the firing of neurons creates consciousness. Nevertheless, the marvelously entertaining books of neurologist Oliver Sacks show in fascinating and sometimes disturbing detail just how intimately our deepest thoughts and feelings, indeed our whole conception of ourselves and our world, are related to brain function. Apparently, if you change your brain you change your *self*. Religion might also clash with the social and human sciences. It is a central tenet of Christian belief that humans are sinners and that sin is a matter of the conscious choice of morally responsible agents. Yet some psychological theories deny that humans have such freedom and interpret human behavior as caused by, for instance, conditioning (behaviorism) or subconscious motivations (psychoanalysis).

Of course, the most famous conflict between science and religion, one that occasionally erupts even in the present day, is the clash between conservative Christianity and evolutionary theory. As historian James Moore showed, it is false that Christian theologians consistently opposed Darwinian evolution from the beginning. As Moore indicates, many Christians were quickly reconciled to Darwinism and even embraced it enthusiastically. Today most Christian denominations officially accept evolution. Pope John Paul II stated that evolution is "more than just a theory," and that it is clearly the correct account of the origin of biological species. Still, many conservative Christians have simply never been able to accept evolution or square it with beliefs that are for them essential elements of Christian doctrine. Perhaps the main problem is that evolution contradicts a straightforward reading of the creation accounts in the Book of Genesis. For today's fundamentalists, like the Young Earth Creationists, this is undoubtedly the basis for much of their animus against evolution.

Not all opponents of evolution are fundamentalists. Phillip E. Johnson, a professor in the School of Law at the University of California at Berkeley, is perhaps the best-known current critic of Darwinism. Johnson is not a Young Earth Creationist. He is perfectly willing to admit that the earth is billions of years old, just as geologists and evolutionary biologists claim. Further, he is not committed to a view of Scripture as inerrant, so it is not the conflict of evolution with a literal reading of Genesis that bothers him. He does think that the evidence for evolution is shoddy, and he has written extensively trying to show that this is so. What really bothers him, though, is not evolution itself but what he sees as an even deeper problem with the reigning assumptions and practice of science. Why, he asks, if the evidence for evolution is so weak, is it so nearly universally accepted among scientists? His answer is that science has been adulterated by a philosophical dogma, the doctrine of metaphysical naturalism. Johnson explains metaphysical naturalism as follows:

> Naturalism assumes the entire realm of nature to be a closed system of material causes and effects, which cannot be influenced by anything from "outside." Naturalism does not explicitly deny the mere existence of God, but it does deny that a supernatural being could in any way influence natural events, such as evolution, or communicate with material creatures like ourselves. (Johnson, 1991, pp. 114–115)

In other words, metaphysical naturalism assumes that all natural things have only natural causes and therefore rejects out of hand any hypotheses postulating supernatural causes, like a divine Creator. It is this *philosophical* bias against the supernatural, says Johnson, not anything necessary for good science, that leads scientists to accept evolution and reject creationism:

> Creationists are disqualified from making a positive case, because science by definition is based upon naturalism. The rules of science also disqualify any purely negative argumentation designed to dilute the persuasiveness of evolution. Creationism is thus ruled out of court—and out of classrooms—before any consideration of evidence. (Johnson, 2001, p. 67)

So, Johnson's argument is that science has compromised its objectivity by ruling out supernatural hypotheses, like creationism, without a hearing while accepting naturalistic theories, like evolution, that have little going for them. How good is Johnson's case? This is not the place to enter into the evidence for evolution. Many good books have done that already. Also, this is not the place to rehash the whole creation/evolution debate. Many fine books, a few of which are listed below in the "Further Readings" section, have done that job admirably. Here we shall address just three questions: (1) Is philosophy ever relevant to the evaluation of scientific hypotheses? (2) Does Science assume metaphysical naturalism? (3) Are supernatural hypotheses like creationism dismissed by philosophical fiat and without a thorough empirical evaluation?

Well, why should there be any *philosophical* debate about *scientific* hypotheses? Why not just run every proposed hypothesis through a good empirical test? After all, isn't the whole point of scientific method supposed to be that we can test hypotheses rather than engage in long-winded philosophical debate? The simple fact of the matter is that far too many hypotheses can be thought up than can possibly be tested. The empirical evaluation of hypotheses is an exacting, time-consuming process that requires meticulous planning and frequently involves the use of very expensive equipment that is often available only on a very competitive basis. Astronomers sometimes have to wait months to get one night on one of the big telescopes, and if it turns out cloudy on their night to observe—too bad. Besides, scientists are very, very busy people. So, before scientists can even begin to consider a hypothesis for testing, it has to show considerable promise. Scientists get ideas all the time, the vast majority of them bad. Like the White Queen in *Through the Looking Glass,* scientists can often think of six impossible things before breakfast. How do we distinguish the hypotheses with promise, the ones that we actually will consider testing, from those that are throwaways?

Partly, scientists judge on the basis of track record. If a new hypothesis is just a variant of a kind that has been tried and has failed repeatedly, scientists are likely to give it short shrift. Now this may be unfair in many cases; some worthy hypotheses may be judged guilty by association. After all, *no* sort of hypothesis has a good track record until one of that sort actually does succeed. But nobody ever said science had to be completely fair, and there just does not seem to be any other way to proceed.

Philosophical considerations also inform our judgments about whether a hypothesis shows promise of test-worthiness. For instance, until well into the twentieth century, many professional biologists advocated *vitalism.* Vitalists held that life and living processes could not be completely explained in terms of the laws of chemistry and physics, so they postulated an additional "vital force" or animating "principle" that was supposed to permeate every tissue of living things. Vitalism was not a silly or obscurantist doctrine. Many of the leading figures in the history of the life sciences, including Pasteur, advocated some form of vitalism. However, vitalism was eventually abandoned in part because it lost repeatedly when placed head-to-head against mechanistic hypotheses. Another reason it was abandoned was that biologists came to see "vital force" as an explanatory dead end. Instead of explaining organic phenomena, invoking "vital force" just seemed to deepen the mystery.

The question of what constitutes a legitimate scientific explanation is a philosophical question, one pursued at considerable length by philosophers of science. Because this is such an important question in the philosophy of science, let's digress a bit to review briefly what some philosophers have said about it. It is widely agreed that one goal of science is to explain the observed features of the physical universe. Science therefore asks questions like these: Why are certain zones of the earth's crust particularly susceptible to earthquakes while others hardly ever have even a tremor? Why are galaxies arranged in clusters and superclusters rather than just spread randomly through space? Why are island faunas so unique, often displaying a range of adaptations not found in related faunas of the nearest mainlands? Scientific explanations relieve our puzzlement about such questions by showing why these particular phenomena were to be expected.

The classic modern model of scientific explanation was articulated by philosophers Carl Hempel and Paul Oppenheim in 1948. Hempel later devoted much effort to refining and extending this model. The basic idea is that what sets scientific explanation apart from other ways of achieving elucidation or enlightenment is that scientific explanations have a distinct *form*. A scientific explanation has the form of an *argument* where a conclusion is drawn from a set of premises. The datum to be explained—what philosophers call the *explanandum*—is scientifically explained when it is correctly inferred from particular kinds of premises. For Hempel and Oppenheim, at least one premise needs to state a *natural law*. Another premise states *initial conditions,* i.e., an appropriate set of concrete physical circumstances. Propositions that state natural laws are called *nomological propositions* (from the Greek word "nomos" meaning "law"). A *Deductive Nomological* (DN) model of explanation is therefore one in which an explanandum is explained by deducing it from premises that state a natural law and a set of initial conditions. Because natural laws play a vital role in this model of explanation, it is called a "covering law" model.

According to Hempel and Oppenheim, many explanations in science have a DN form. Here is a simple example:

> Natural law: When water freezes it expands with enormous force.
> Initial conditions: The water in the pipes froze solid overnight.
> Conclusion: The pipes burst.

Given that freezing water exerts an enormous expansive force—a force too great for household plumbing to contain—and given that the water in our pipes did freeze last night, we can *deduce* that our pipes burst. So, if we want to know why we have burst pipes, and a terrible plumber's bill to pay, we gain scientific understanding (but not much solace) when we know that freezing water expands and that the water in our pipes froze.

Many explanations in science fit the DN model and the allied *Inductive Nomological* (IN) model, which is much the same as the DN model, except that we infer the *probable* occurrence of the explanandum from a law and initial conditions. However, as various philosophers have pointed out, not all legitimate scientific explanations conform to the DN or IN models. Here is an example of one that does not:

> Any unvaccinated person exposed to live influenza virus has a 20 percent to 40 percent chance of developing a case of the flu within 72 hours. Sam

has had no flu shot this year and he sat for two hours in a movie theater next to Sarah, who was just coming down with an active case of the flu. Two days later Sam developed a case of the flu. Therefore, Sam got the flu because he was exposed to live influenza virus he got from Sarah.

Surely this is a legitimate explanation of why Sam got the flu, but it fits neither the DN nor the IN model. You cannot *deduce* that Sam will get the flu from the fact that he is unvaccinated and has been exposed to flu virus. It is not *certain* that Sam will get the flu; there is only a 20 percent to 40 percent chance. Also this explanation does not fit the IN model because it is not even *probable* that Sam will get the flu. If there is a 20 percent to 40 percent chance that Sam will get the flu, there is a 60 percent to 80 percent chance that he will not. So it is *probable* that Sam will *not* get the flu despite his exposure to the virus. Still, if Sam *does* get the flu, exposure to the virus is the explanation.

To deal with cases like this, philosophers developed the *Causal Statistical* (CS) model of scientific explanation. According to the CS model, we understand an event when we spell out the physical factors statistically relevant to the event and also specify the underlying causal processes that brought about the event. For instance we above explained why Sam came down with influenza by noting that he is unvaccinated and he came in contact with another person with an active case. To say that exposure to influenza is statistically relevant to Sam's getting the flu does not mean that such exposure makes it *likely* (i.e., more than 50 percent probable) that Sam will get the disease. It only means that such exposure makes it *more likely* that Sam will get the flu than if he had not been exposed. A 20 percent chance of the flu is greater than a 0 percent chance. We expand our explanation by specifying just how the influenza virus does its dirty work on the body. Viruses display a malign ingenuity in the way that they invade body cells and hijack the cell's genetic machinery to make more copies of themselves. Such knowledge about how viruses operate greatly expands our understanding of the facts about infection.

Other philosophers reject all such models of scientific explanation and recommend a pragmatic approach. According to these philosophers, all we can really say about a good explanation is that it answers our "why" questions about a given topic of concern by telling why *this* particular outcome was to be expected rather than one of the other members of that event's "contrast class." The "contrast class" consists of all of the other events that conceivably could have occurred in that situation but did not. For instance, a satisfactory scientific explanation of the dinosaurs' demise would tell us why the dinosaurs went extinct, but crocodilians sailed right through the K/T mass extinction and are with us today.

Let's return from this (all too brief) review of what some philosophers have said about scientific explanation to the main question: Are philosophical considerations ever relevant to theory choice? In particular, might a philosophical consideration, like our ideas about what constitutes a good scientific explanation, reasonably guide us in deciding which hypotheses are promising candidates for further testing and which are nonstarters? It is important to note that concern about the nature of scientific explanation is not merely an armchair amusement for philosophers. As we saw in Chapter Two, Kuhn notes that scientists themselves often engage in vigorous debates over standards, like what should constitute a legitimate explanation for some range of natural phenomena. It seems therefore that philosophical considerations should sometimes guide us in choosing which hypotheses look promising enough to go to the trouble of testing. We cannot test every hypothesis that anyone proposes. If a hypothesis looks like it does not even offer us a good explanation, it is not unreasonable or unfair to pass it over, at least for the time being, in favor of something more promising.

Well what about supernatural hypotheses? Does philosophical bias prevent them from receiving due consideration? One complaint often made against hypotheses that invoke God's acts is that they do not explain things, but only hide our ignorance behind a theological fig leaf. For skeptics, saying "God did it" does not enhance our understanding of some strange phenomenon—a sudden, unexplained remission of metastatic cancer, for instance—but only drapes it in deeper mystery. Is this accusation fair? Do hypotheses that invoke God, or perhaps a more nebulous Creator, offer legitimate explanations, or are they only markers for our ignorance, placeholders for explanations we hope someday to get?

Defenders of supernatural hypotheses could strengthen their case if they could show that their hypotheses offer explanations that conform to one or more of the recognized models of scientific explanation. Let's consider whether there could be supernatural explanations that conform, for instance, to the DN or CS models. There do not seem to be any "laws of supernature" to serve as covering laws to explain particular events, so the DN model seems to be out. For instance, we just do not know the general circumstances in which God is likely to perform a miracle. We cannot articulate any general laws of the form "Every time God's people are in dire enough need, God performs a public miracle to deliver them." Putative beings like gods and ghosts are not constrained by natural law; their actions are unpredictable so it is hard to know what effects of those actions are to be expected. Whether supernatural hypotheses specify statistically relevant factors for the occurrence of events is a matter of debate among philosophers of religion. Some theistic philosophers argue that the existence of the universe, or of a particular kind of universe, is more likely if there is a God than if there is not. However, nobody can specify the particular causal processes whereby God is supposed to bring about his effects; after all, God's ways are proverbially mysterious. Nobody can be much more specific than to say that when God created something—birds, let's say—he just said "Let there be birds" and POOF! there were birds! Supernatural hypotheses are like the famous Sidney Harris cartoon showing two scientists standing before a chalkboard full of mathematical scribbles except in the middle where it says, "step two: A miracle occurs." One scientist comments dryly "I think you need to be a little more specific here in step two." This lack of specificity about causal mechanisms means that supernatural explanations do not conform to the CS model.

Supernatural explanations do not even meet our pragmatic explanatory needs very well. For instance, if we try to explain the sorts of anatomical homologies mentioned in the last chapter by saying that God created organisms according to a plan, this leaves all of our questions unanswered. Why this plan rather than one of the indefinitely many others that an all-powerful, all-knowing creator could have enacted? Why just this instance rather than one of an indefinitely large contrast-class?

Defenders of supernatural hypotheses will counter, correctly I think, that the models of scientific explanation so far developed do not necessarily exhaust all the legitimate possibilities. While philosophers of science may have identified *some* kinds of good scientific explanation, it is highly questionable whether they have identified *all* possible types. Therefore, the fact that supernatural explanations do not conform to any "model of scientific explanation" so far proposed does not mean that they cannot be legitimate scientific explanations. Fair enough, but surely the burden of proof is on the defenders of supernatural hypotheses. *Prima facie* such hypotheses do not seem to offer much elucidation. Again, if we say, for instance, that God created birds, what does that tell us? How did He do it? For what reason? Why birds? Why didn't He stick with the highly successful flying reptiles? Surely, any kind of acceptable scientific explanation should show why the explanandum—the existence of birds in this case—was to be expected. As philosophers Karel Lambert and Gordon G. Brittan, Jr. note (1987, p. 22), invocations of God's will, like appeal to signs of the Zodiac, just do not provide such information.

The upshot is that there *is* a philosophical motivation behind the scientific practice of giving short shrift to supernatural hypotheses, just as Johnson says. But until defenders of supernatural hypotheses can show that such hypotheses promise legitimate scientific explanations—and do not just disguise our ignorance—such practice is neither biased nor unfair. Please note that this does *not* mean that supernatural hypotheses cannot be true or that we cannot have very good reasons for thinking them true (more on this below). Maybe we will just have to admit that some things do not have a scientific explanation and are due to the mysterious acts of a Creator. Maybe on some topics, like the origin of life, say, scientists may someday come to the point where they should just throw up their hands and say that they will never explain some things and concede that there are ultimate mysteries, facts attributable only to the unfathomable and inscrutable actions of a Creator. But it is far from clear that that day is today.

Let's move to the second question raised by Johnson's critique: *Does science assume metaphysical naturalism?* This charge has been made many times, and the standard reply is that the naturalism science assumes is methodological, not metaphysical. The difference is this: Metaphysical naturalism is a doctrine about the nature of reality. It can take the strong line that only natural things are real or the weaker line that supernatural things might exist but they cannot causally interact with the natural world. Methodological naturalism does not offer opinions about the nature of ultimate reality; it merely requires that, as a matter of good scientific practice, we consider only naturalistic hypotheses. T. H. Huxley, Victorian scientist and man of letters, was very emphatic that metaphysical questions about the nature of ultimate reality were none of the business of science. Huxley said that you might as well inquire into the politics of extraterrestrials as to ask whether ultimate reality is material or spiritual. Yet he strongly advocated naturalism as a methodological requirement because he held that naturalistic explanations are comprehensible while supernatural explanations only hide mysteries behind a veil of theological obscurity. In a similar vein, contemporary philosopher Rob Pennock argues that science should be godless in the same sense that plumbing is godless. Good plumbing practice obviously does not involve grandiose metaphysical assumptions, but proceeds on the assumption that the cause of a problem is in the pipes. Pennock argues that the requirement that scientific hypotheses be testable entails that they involve only natural objects that follow predictable laws. As noted earlier, putative supernatural entities, like gods and ghosts, are not bound by natural law, and so are notoriously difficult to test.

Johnson does not buy these arguments and insists that methodological naturalism is only a dishonest front for metaphysical naturalism. I think we should concede that *in principle* good science could confirm supernatural hypotheses, however difficult they might be to test *in practice*. Nineteenth-century English scientist Francis Galton proposed a test for the efficacy of prayer. He noted that members of the royal family certainly were the beneficiaries of more prayers for their health than any other British family. He concluded that, if prayer works, the royal family should be healthier than other comparable families (he found that they were not healthier, by the way). Now a legitimate test of the efficacy of prayer is probably impossible to achieve in practice. How could Galton rule out that many disgruntled people may have been praying that God strike down the royal family? Still, this seems to be a practical difficulty, and not an indication that an experimental test of prayer is in principle impossible.

Interestingly, the Bible tells of an incident that would be about as good an experimental test of God's power as anyone could devise. I Kings, chapter 18, tells the story of Elijah and the priests of Baal. Elijah challenged the priests of Baal to a contest to see which god was real, Baal or the Lord, the God of Abraham, Isaac, and Israel. The priests of Baal built an altar and placed a sacrifice upon it. All day they cried for Baal to send fire to burn their sacrifice, but nothing happened. At the day's end Elijah erected an altar, placed a sacrifice on it, and had everything thoroughly soaked with water. He then called upon the Lord, and according to I Kings 18:38, "Then the fire of the Lord fell and consumed the burnt offering and the wood and the stones, and the dust, and licked up the water that was in the trench." Now this would certainly seem to be about as good an example of a crucial experiment as any scientist has ever devised. If it occurred today, the churches and synagogues would fill with former doubters. Of course, such things apparently do not happen today, but the point is, again, that there seems to be nothing *in principle* impossible about an experimental test of God's power.

So, is it simply a matter of ideological prejudice that supernatural hypotheses are rejected by science? No, for two reasons (besides the doubts raised earlier about supernatural "explanations"): First, though it is not a methodological *requirement* of science, naturalism has unquestionably proven a valuable *heuristic*. A "heuristic" is a presumption that serves as a guide for inquiry. An example of a heuristic principle that guides science is the principle of simplicity, the postulation that physical reality is ultimately simple, and that science should therefore seek simple theories. The idea that things will ultimately turn out to be simple is, of course, a speculation. Absolutely nothing guarantees that at bottom physical reality is simple. Yet no one can deny that the presumption that deep simplicity underlies the surface complexity of nature has been an extremely valuable heuristic guiding science.

Similarly, naturalism has been a very successful heuristic principle. Unquestionably, much of the progress of science is due to the fact that it doggedly sought natural hypotheses and excluded those postulating gods, souls, angels, demons, ghosts, fate, magic, astrological influences, hexes, spells, good luck charms, and so forth. So long as a heuristic continues to deliver the goods, scientists are fully justified in sticking with it. Is there any indication that a naturalistic heuristic has served its purpose and now leads science in the wrong direction? For instance, does naturalism induce scientists to accept evolution despite a dearth of evidence? For decades, anti-evolutionists have charged that evolution is a "theory in crisis" and that Darwin is once again "on trial." They have insisted repeatedly that the evidence for evolution is so shoddy that the whole edifice of evolutionary science is about to come crashing down and that the only thing propping it up is naturalistic bias.

Let's pause for a second and consider just how strong a claim this is. A recent thorough electronic search of the professional, peer-reviewed scientific literature over the previous twelve-year period turned up over 100,000 articles with "evolution" as a key word and, by the way, practically none referring to concepts of supernatural design. So, if evolutionary theory has been "in crisis" and "on trial" for decades, this news has yet to reach the writers of the professional scientific literature. Evolutionary biology looks extremely spry for a field supposedly on its deathbed! Johnson and other anti-evolutionists have to attribute evolutionary biology's appearance of health and vigor to a massive intellectual fraud perpetrated on science by a cadre of ideologues. But no ideology, not even when backed by the power to burn dissenters at the stake, has ever held science down for long. Not even the enormous power and intellectual influence of the seventeenth-century Church could hide the bankruptcy of the old Ptolemaic system for long. So, it is just hard to believe that nothing but ideological obscurantism keeps scientists from recognizing the alleged weakness of evolutionary theory.

A second and more important reason for denying that negative attitudes towards supernatural hypotheses are due to bias is that Johnson's charge is simply false. The answer to the third of the questions we are addressing in this section is: No, creationism has not been dismissed by philosophical fiat. Dozens of books and hundreds of articles, many available on the Internet in the magnificent talk.origins archives, have subjected creationist claims to careful, extensive, point-by-point empirical critique. In *The Origin of Species* Darwin himself showed time and time again that natural selection better explains the natural facts than special creation. Therefore, the creationist hypotheses have not been rejected by philosophical fiat and without a fair and thorough hearing.

CONCLUSION

In this chapter we have extensively examined critiques of scientific rationality and objectivity from both the left and the right. Our conclusion has to be that, though science is far from perfect, as any human enterprise must be, there is still something left of the Enlightenment ideal derided by Latour. There is a physical world "out there," and we can know some things about it. That is, we can say of the natural world, without qualification or apology, that some things really just *are* so, and are not artifacts of our percepts, concepts, or categories. Further, our observations of the physical world can be used to rigorously evaluate our theories, so that our theoretical beliefs are shaped and constrained by nature, and not merely by politics, rhetorical manipulation, or ideology. Disinterested knowledge really is possible, and is in fact achieved far more often than cynics suppose.

Yet even their harshest critics must admit that the social constructivist, postmodernist, and feminist science critics have performed a valuable service. These critics have certainly succeeded in disposing of what might be called the "passive spectator" stereotype of scientific knowledge. According to this stereotype, modern science began when the pioneers like Copernicus, Galileo, and Darwin stopped bowing to ancient authority and opened their eyes to the world around them. Once people started looking at *nature* rather than old books, scientific knowledge flowed into open scientific minds like water pouring into an empty bucket. Now, of course, this is a comic-book version of the history of science, and no serious scholar has ever thought that it really happened this way or that we gain scientific knowledge merely by the passive reception of information. Still, this has been a very influential stereotype. A powerful image can influence our thinking more than all the careful arguments of scholars. One of the indelible images of our intellectual culture is the picture of Galileo boldly scanning the heavens with his telescope, eager to discover whatever his eyes revealed to him, while his ecclesiastical oppressors, besotted with Scripture and Aristotle, refused even to look through the instrument. The founders of Britain's Royal Society, the preeminent British scientific body, were so impressed with this image of the scientist as the ideally objective and open-minded observer that they adopted as the Society's motto "*Nullius in Verba.*" This motto is hard to translate precisely, but it means that you should take nothing on authority. Instead, you should look and see for yourself.

Scientific discovery requires active engagement, however, not just passive seeing. Galileo didn't just look through the telescope and report what he saw; he interpreted, theorized, speculated, measured, analyzed, and argued. Darwin did not just go to the Galapagos Islands, see some odd finches and tortoises, and then awaken to the truth of evolution in a flash of blindingly obvious insight (scientific discoveries are always "obvious" only in hindsight). Darwin's private notebooks, written as he struggled to define his ideas on evolution, reveal a complex process of questioning, argument, and counterargument, with tentative conclusions drawn and then repeatedly rejected or refined. Scientists do not just *absorb* a picture of the world; they *create* a picture and then do their best to see how accurate it is. Unavoidably, when we create our theories of the natural world, we must employ the only cognitive tools we have—the concepts, language, perspectives, interpretive and observational skills, and presumed background knowledge that we possess. Inevitably, multifarious biases lurking in our language and concepts will sometimes—all too often—slip unnoticed into our theories. Our only way of dealing with this problem is to continually refine and revise our ideas through ongoing interaction with the natural world and the effort to devise stricter methods, more rigorous tests, and more accurate measures. The work of Kuhn examined in the last chapter, and that of the radical science critics considered in this one, unquestionably succeeded in debunking the simplistic stereotype of the scientist as ideally objective, open minded, and a passive observer.

What we need then is a balanced view of science, one that rejects both the excessively cynical and the unrealistically idealized stereotypes of science. David Young, in his excellent book *The Discovery of Evolution,* strikes just the right note of balance in our interpretation of science; his words can serve as a coda for this chapter:

> The picture of the scientist as an objective spectator has died a natural death, thanks to the work of historians and philosophers of science. It is now clear that even simple observations are not imbibed passively from the external world but are made by a human mind already laden with ideas. The shaping of these ideas is a human activity carried out in a particular social context, with all the frailties and limitations that that implies. This has led some people to the other extreme, in which scientific knowledge is viewed as no more than the expression of a particular social group. On this view there are no such things as discoveries in science, only changes in fashion about how we choose to view the world. However, such a view cannot account for the fact that scientific understanding does not merely change but is progressive. . . . A sensible view of scientific theory must lie somewhere between these two extremes and embody elements of both. Certainly, scientific discovery does not involve a one-way flow of information from nature to a passive, open mind. It involves a creative interaction of mind and nature, in which scientists seek to construct an adequate picture from what they see of the world. (Young, 1992, pp. 219–220)

Young's view is neither novel nor especially profound. It lacks all of the edgy excitement of the radical science critiques. It only has one big advantage over those accounts: It is true.

FURTHER READINGS FOR CHAPTER THREE

Paul de Kruif's *Microbe Hunters* (San Diego: Harcourt Brace & Company, 1926), is still in print eighty years after it was written. I remember being fascinated as a child reading an old dog-eared paperback copy. I am sure that it has inspired many readers to enter medicine or biomedical research. As I say in this chapter, de Kruif regards Pasteur and the other microbe hunters with unabashed hero worship and he treats the pursuit of science as the noblest and most selfless of activities. We now, of course, realize that these views are naïve. Ambrose Bierce once defined a saint as "a dead sinner, revised and edited," and no doubt well-meaning admirers like de Kruif have likewise redacted the stories of the "saints" of science. Historians of science perform a valuable service when present the story "warts and all." However, "warts and all" does not mean "nothing but warts." In my view, undercutting one myth, the myth of the saintly scientist grappling with the demons of disease, should not lead to the creation of more pernicious stereotypes, such as the image of the scientist as cynical self-promoter, rampant ideologue, or stooge of vested interests.

Upon reading a draft of this chapter, one referee said that I had introduced Bruno Latour as the "villain" who insulted the memory of France's national hero of science, Pasteur. No. I present Latour as a radical revisionist, a characterization I have no reason to think he would repudiate. In fact, Latour boasts of the deflationary intention of his work. In a letter to the editor of the (now, sadly, defunct) magazine *The Sciences* (vol. 35, no. 2, p. 7, 1995), Latour compares his work in science studies to the work of Darwin in biology: "Those of us who pursue science studies are the Darwins of science, showing how the exquisite beauty of facts, theories, instruments and machines can be accounted for without ever resorting to teleological principles or arguments by design." I see no other way to read this passage than as a statement of Latour's intent to replace traditional representations of science as motivated by *reasons* with a reductionistic sociological analysis. And that is how I have presented him in this chapter.

There is a lot to be said for the slogan "it is more important that an opinion be interesting than true." Some errors are uninteresting because they are due to silly mistakes in reasoning; others are interesting because they involve deep confusions in our concepts or language. Bruno Latour's errors are *never* dull. When he is wrong, you learn a lot about science and how it operates in society even as you try to pinpoint his errors and sort out what is really behind his conclusions. Latour made his big splash with *Laboratory Life: The Construction of Scientific Facts* (Princeton: Princeton University Press, 1979; reprinted with new postscript and index in 1986). This book, co-written with Steve Woolgar, was one of the founding documents of the whole "science studies" movement. It is rather technical in places, and the prose is often muddy. Still, the aim of going into a laboratory as an anthropologist, to observe scientists in their native habitat as one would the Yanomamo or the Inuit was a brilliant idea and makes for fascinating reading.

Latour examines Pasteur and his influence in *The Pasteurization of France* (Cambridge: Harvard University Press, 1988), translated by Alan Sheridan and John Law. Latour's most ambitious and comprehensive work is *Science in Action: How to Follow Scientists and Engineers through Society* (Cambridge: Harvard University Press, 1987). Latour's thought took an interesting turn in 1993 with the publication of *We Have Never Been Modern* (Cambridge: Harvard University Press), translated by Catherine Porter. Here Latour claims to abandon social constructivism and aims to explore a middle course between constructivism, the view that scientific facts are cultural artifacts, and the view of scientists that such objects are objective truths about nature. Latour defines what he calls "quasi-objects" that are neither wholly natural nor wholly constructed. Unfortunately, just what he means by a "quasi-object" is not made entirely clear.

Steve Woolgar develops his views in confrontational style in *Science: The Very Idea* (London: Tavistock Publications, 1988). One very well-known scientist and writer who, at least sometimes, seemed to endorse social constructivism was Stephen Jay Gould. His book *The Mismeasure of Man* (New York: W. W. Norton, 1981) is often cited by social constructivists, postmodernists, and feminists, as proof of how bias and bigotry shape science. Matt Cartmill's perceptive but acerbic characterization of Latour's social constructivism is found in his review of *Mystery of Mysteries: Is Evolution a Social Construct?* by Michael Ruse. This review appeared in *Reports of the National Center for Science Education* 19, no. 5 (1999), 49–50. The story of the wrongheaded dinosaur episode is found in Chapter One of my book *Drawing Out Leviathan* (Bloomington, Ind.: Indiana University Press, 2001). Michele Marsonet's very interesting discussion of the way that philosophy can slide into "linguistic idealism" is found in his *Science, Reality, and Language* (Albany, N.Y.: State University of New York Press, 1995). John Searle's discussion of the "default settings" of the human mind are found in his book *Mind, Language, and Society* (New York: Basic Books, 1998).

As I say, the roots of postmodernism can be traced at least back to Nietzsche. Nietzsche is an exciting but challenging thinker. His writings often have a declamatory or oracular character, which puts some readers off. Also, he is very easy to misread. Sometimes he sounds like an anti-Semite, a misogynist, or a proto-fascist, though his defenders insist that he was none of these things. Because of some of the difficulties with reading Nietzsche, it might be good to start by reading a reliable introduction to his thought. A good, succinct, and readable account is *On Nietzsche,* by Eric Steinhart (Belmont, Calif.: Wadsworth, 2000).

An essential document for understanding postmodernism is Lyotard's manifesto, *The Postmodern Condition: A Report on Knowledge* (Manchester: Manchester University Press, 1979), translated by Geoff Bennington and Brian Massumi. Richard Rorty's *Philosophy and the Mirror of Nature* (Princeton: Princeton University Press, 1979), introduced many characteristic postmodernist theses to English-speaking philosophers. Rorty's work attracted quite a bit of notoriety because he had previously been regarded as one of the really tough-minded philosophers of the analytical tradition, and it seemed to many of his contemporaries that he was simply abandoning philosophy. A fun and very accessible introduction to postmodernism is Glen Ward's *Postmodernism* (NTC/Contemporary Publishing, 1997). Postmodernism is largely a development of recent literary theory, and a perceptive critique of postmodernism by an expert on such theory is *The Illusions of Postmodernism* by Terry Eagleton (Oxford: Blackwell Publishers, 1996).

The two works I selected to represent postmodernist commentary on science were Donna Haraway's *Primate Visions: Gender, Race, and Nature in the World of Modern Science* (New York: Routledge, 1989) and W. J. T. Mitchell's *The Last Dinosaur Book: The Life and Times of a Cultural Icon* (Chicago: University of Chicago Press, 1998). What makes these books particularly interesting is that they deal with primatology and dinosaur paleontology, which are branches of science that are easier for most people to relate to than, say, particle physics. When scientists encounter Haraway's and Mitchell's books, they often are nonplussed or outraged. Matt Cartmill's trenchant review of *Primate Visions,* published in the *International Journal of Primatology* 12, no. 1 (1991), must surely express the exasperation many primatologists would feel toward Haraway's book. I coauthored an essay with geologist Peter Copeland, "Toward a Postmodernist Paleontology?" in *Academic Questions* 17, no. 2 (spring 2004) that examines and criticizes Mitchell's claims in detail. Alan Sokal's *faux*-postmodernist essay, "Transgressing the Boundaries: Towards a Transformative Hermeneutics of Quantum Gravity," is most conveniently found in *The Sokal Hoax: The Sham That Shook the Academy* (Lincoln, Neb.: University of Nebraska Press, 2000), edited by the editors of the magazine *Lingua*

Franca. This volume is a lot of fun, with fierce polemics and much out-raged harrumphing on both sides.

A good place to start with the feminist philosophy of science is the entry "Feminist Accounts of Science" by Kathleen Okruhlik, in *A Companion to the Philosophy of Science,* edited by W. H. Newton-Smith (Oxford: Blackwell Publishers, 2000). Okruhlik provides an authoritative overview of the diverse views of feminist philosophers of science. As I state in both Prefaces, Sandra Harding appears in this chapter not as a "typical" representative of feminist philosophy of science (though I do not consider her too atypical either) but as a controversial figure whose opinions are bound to excite discussion. If Harding has piqued your interest in feminist philosophy of science, you should probably next read Elizabeth Fox Keller's *Reflections on Gender and Science* (New Haven: Yale University Press, 1985), and Helen Longino's *Science as Social Knowledge: Values and Objectivity in Scientific Inquiry* (Princeton: Princeton University Press, 1990). The Harding quote in this section comes from her *Whose Science? Whose Knowledge? Thinking From Women's Lives* (Ithaca, N.Y.: Cornell University Press, 1991). For convenience, I quoted Harding from the selection "Feminist Standpoint Epistemology and Strong Objectivity" from the book *The Science Wars* (bibliographical details below).

For critiques of the feminist philosophy of science, see Cassandra Pinnick, "Feminist Epistemology: Implications for the Philosophy of Science," in the journal *Philosophy of Science* 61 (1994): 646–657; Janet Radcliffe Richards's "Why Feminist Epistemology Isn't," and Noretta Koertge's "Feminist Epistemology: Stalking an Un-Dead Horse." Richards's and Koertge's essays are found on pages 385–412 and 413–419, respectively, in Paul Gross, Norman Levitt, and Martin Lewis, eds., *The Flight From Science and Reason* (New York: New York Academy of Sciences, 1996). A book-length critique of feminist epistemology and philosophy of science is Ellen R. Klein's *Feminism Under Fire* (Amherst, N.Y.: Prometheus Books, 1996). Steven Pinker's provocative discussion of gender differences is found in his *The Blank Slate: The Modern Denial of Human Nature* (New York: Viking, 2002). The quote from John Stuart Mill is from his classic *On Liberty,* in *Great Books of the Western World,* Robert Maynard Hutchins, editor-in-chief (Chicago: Encyclopedia Brittanica, Inc., 1952).

The debates over social constructivist, postmodernist, and feminist accounts of science reached the boiling point in the mid-1990s as scientists began to fire back at what they perceived as attacks on the aims, methods, and values of science by critics of the "academic left." The "science wars" really erupted with the 1994 publication of Paul Gross and Norman Levitt's splendidly pugnacious *Higher Superstition: The Academic Left and its*

Quarrels with Science (Baltimore: Johns Hopkins University Press). The academic left responded in equally bellicose fashion in the collection of essays *Science Wars,* edited by Andrew Ross (Durham, N.C.: Duke University Press, 1996). Alan Sokal, of the Sokal hoax, and his collaborator Jean Bricmont stated their case in *Fashionable Nonsense: Postmodern Intellectuals' Abuse of Science* (New York: Picador USA, 1998). A good collection of critiques of the left-wing science critique is *A House Built on Sand: Exposing Postmodernist Myths about Science,* edited by Noretta Koertge (Oxford: Oxford University Press, 1998). By 2000, the rhetorical temperature of the science wars had cooled a bit and books appeared that were less polemical in tone. A clear and insightful survey of the main issues debated in the science wars in the context of the recent history of the philosophy of science is James Robert Brown's *Who Rules in Science: An Opinionated Guide to the Wars* (Cambridge: Harvard University Press, 2001). I offer an introduction to some of the main writings and debates in the anthology *The Science Wars: Debating Scientific Knowledge and Technology,* edited by Keith M. Parsons (Amherst, N.Y.: Prometheus Books, 2003).

Stephen Jay Gould's views on the relation between science and religion are found in his book *Rock of Ages: Science and Religion in the Fullness of Life* (New York: Ballantine Publishing, 1999). A solid, thorough overview of the relations between science and religion is John Hedley Brooke's *Science and Religion: Some Historical Perspectives* (Cambridge: Cambridge University Press, 1991). A detailed yet quite readable examination of the theological response to Darwinism in Britain and America is James R. Moore's *The Post-Darwinian Controversies: A Study of the Protestant Struggle to Come to Terms with Darwin in Great Britain and America, 1870–1900* (Cambridge: Cambridge University Press, 1979).

The recent controversy over "intelligent design," actually a continuation of the controversy over creationism of the 1980s, was kicked off by the publication of Phillip E. Johnson's *Darwin On Trial* (Washington, D.C.: Regnery Gateway, 1991). A history and overview of the intelligent design movement written by a sympathizer is Thomas Woodward's *Doubts about Darwin: A History of Intelligent Design* (Grand Rapids, Mich.: Baker Books, 2003). An anthology of writings by advocates of intelligent design theory with responses by critics is *Intelligent Design Creationism and its Critics: Philosophical, Theological, and Scientific Perspectives* (Cambridge: MIT Press, 2001, ed. by Rob Pennock), which includes the essay by Johnson quoted in this chapter, "Evolution as Dogma: The Establishment of Naturalism."

The volume of literature on evolution is simply stupendous in its quantity (and highly variable in its quality). Here I shall simply recommend one book that seems to me the best presentation of evolutionary theory for the nonspecialist, Colin Patterson's *Evolution,* second edition (make sure you get the second edition, it is much better than the first) (Ithaca, N.Y.: Comstock Publishing, 1999). Patterson's treatment is crystal clear, and, while it presents the evidence for evolution cogently, it is very undogmatic in tone. Really to understand Darwinism, you need to see it presented in the context of its historical development. A superbly written, insightful, and beautifully illustrated history of evolution is David Young's *The Discovery of Evolution* (Cambridge: Cambridge University Press, 1992). Perhaps the best nontechnical statement of the intelligent design position is still the Johnson book mentioned above. I think that for most readers the best critique of Johnson's view and the claims of intelligent design creationism is Robert T. Pennock's *Tower of Babel: The Evidence Against the New Creationism* (Cambridge: MIT Press, 1999).

One very good introduction to the topic of scientific explanation is the chapter "Scientific Explanation" by Wesley C. Salmon in Salmon, et al., *Introduction to the Philosophy of Science* (Englewood Cliffs, N.J.: Prentice-Hall, 1992). Salmon, one of the top philosophers of science of the twentieth century, made many seminal contributions to our understanding of scientific explanation. He was also a gifted expositor who could make difficult ideas very clear for beginners. Another very clear and helpful overview of the topic of explanation is in *An Introduction to the Philosophy of Science,* third edition, by Karel Lambert and Gordon G. Brittan, Jr. (Atascadero, Cal.: Ridgeview Publishing Company, 1987).

T. H. Huxley's comments on methodological naturalism are found in his essay "On the Physical Basis of Life," in *Selected Works of Thomas H. Huxley* (New York: D. Appleton, no date), 130–165. Huxley's prodigious learning, wit, and trenchant style are as enjoyable now as they must have been discomfiting for his nineteenth-century opponents. "Darwin's bulldog" still has considerable bite. For full details on the search of scientific literature that turned up 100,000 articles with "evolution" as a key word, see Staver, J. R., "Evolution and Intelligent Design." *The Science Teacher* 70, no. 8 (2003): 32–35.

4 🐾

ASCENDING THE SLIPPERY SLOPE
Scientific Progress and Truth

How do the natural sciences differ from other fields of human intellectual and creative endeavor? Most people would say that science progresses in ways that, for example, the visual arts do not. This is not to say that the visual arts do not progress. The discovery of perspective by Renaissance painters gave artists a new technique that they used to create some of the most memorable masterpieces, like Raphael's *School of Athens*. However, it would be rash indeed to claim that the art being produced now is better, in any absolute sense, than the art of the Renaissance masters. Yet we generally have no qualms at all about saying that we have far more scientific knowledge now than we did centuries ago. That is, our scientific beliefs are not just regarded as different, but as *better*. For instance, when the Black Death swept Europe in the middle of the fourteenth century, killing approximately 40 percent of the population, the best "scientific" explanation that the most learned medical authorities could offer was that the catastrophe was caused by a malignant conjunction of the planets Jupiter, Saturn, and Mars. For most people, the cause of the calamity was beyond any scientific understanding and was simply due to the wrath of God. Now we know that the Black Death was an outbreak of the bubonic plague, a highly infectious disease caused by the bacterium *Yersinia Pestis*. The plague is spread by rats that carry fleas infected with the plague germ, and wherever people lived in the fourteenth century there were also rats. When it came to treating the plague, fourteenth-century people were completely helpless. In the present day, the few people who do get bubonic plague can be quickly and effectively treated with antibiotics. So we clearly seem to know more about plague and how to treat it than they did 650 years ago.

How then is it that science is supposed to progress in ways that most other human endeavors do not? Quite simply, science is supposed to progress towards *truth*. Of course, many others, writers for instance, also strive for truth. But it is decidedly *not* the case that writers of the present day are any better at telling us the truth about ourselves than Shakespeare was 400 years ago, or Euripides 2,400 years ago. Yet it is a commonplace that schoolchildren today can tell us things about the universe that Aristotle never dreamed of. In other words, the best writers of today cannot expect to surpass the greatest of the past, but any scientifically literate person of today can give us more accurate information about the nature of the physical universe than the greatest Greek or medieval scientist. At least, this is how the story goes.

THE EVILS OF WHIG HISTORY

The story of science has usually been depicted as a triumphant march from darkness into light. According to this kind of story, we could graph the history of science by letting the horizontal axis of our graph represent the last 10,000 years of human history and the vertical axis the level of scientific knowledge. The line plotting the rise of scientific knowledge would be fairly level and not too far above zero during most of human history. It would bump up a little with the start of agriculture and the invention of the wheel. It would start to rise about 2,500 years ago, slowly at first but steadily. There might be a slight decline during the "dark ages" of Europe when some knowledge was lost, but then an explosive rise starting with the scientific revolution of the seventeenth century. The learning curve would get steeper and steeper in subsequent centuries until, in our day, it is just a few degrees from vertical.

The story of science as an ever-steeper learning curve is an attractive picture. As Latour said in the last chapter, we would like to think that amid the general chaos and confusion of human life there exists an oasis of steady progress. The story of science as a steady march of progress is called "Whig history." Just how it got this name needs some explaining. The Whigs, of course, were the British political party that opposed the Tories. The historian Herbert Butterfield coined the term "Whig history" to describe the kind of history written by Whig partisans. These partisans would always depict the course of English history as a struggle between the progressive and enlightened Whigs and the backwards and benighted

Tories. Naturally, on their account, the historical figures that supported Whig programs were the good guys and the villains were those who stood in the way. Butterfield said that the history of science is told in this Whiggish vein. Those who formulated or anticipated the theories we now accept were lauded as the true scientists while those who took an opposing line are ignored or castigated as obscurantists or pseudoscientists.

The problem with the Whig approach to the history of science is that it is inaccurate and unfair. To graph the history of science as an ever-steeper learning curve, we have to be very selective about the data points we chart. We have to make sure that we chart only those theories and discoveries that we continue to accept as valid. What about all the blind alleys, theories that were quite successful at the time but eventually led nowhere? What about the scientists who went down those blind alleys? Many of them were among the most accomplished scientists of their day. Does the fact that they turned out to be wrong make them "bad guys"? In judging any historical figure it is manifestly unfair to expect him or her to know everything we know today. As the proverb puts it, hindsight is always 20/20. All that anyone can be expected to do is to make the best judgments they can given the information and investigative tools available at a particular time and place. Those learned doctors who attributed the Black Death to the malignant conjunction of planets were not being stupid. Lacking all of the advanced equipment and techniques that biomedical science now enjoys, and lacking the knowledge we have gained in the last 650 years, there was no way that they could have guessed the truth.

As we said in Chapter One, the past is another country with different ways and customs. But the ways they did things in the past seemed as sensible to them as our ways seem to us today. They made just as serious an effort to base their beliefs on the facts and accurate reasoning as we do. Just as it would be ignorant and foolish to laugh at the odd-seeming customs of a rainforest tribe, so it is equally presumptuous to scorn past theories merely because they are outmoded or past scientists because they advocated those theories.

To see how unfair Whig history can be, recall Richard Owen, the scientist whose archetype theory opposed Darwinism. Owen was not a lovable man. Prickly, egotistical, and arrogant, he could be both petty and spiteful in his dealings with other scientists. However, in the eyes of posterity, his anti-Darwinism was his greatest sin. Owen engaged in long and bitter controversy with T. H. Huxley, "Darwin's bulldog." Because Darwin won, Whig historians painted Owen as the arch-obscurantist, a poor scientist who opposed Darwinism on ideological rather than scientific grounds. Owen's criticism of Darwin was particularly rancorous. Some of Darwin's letters reveal that he was deeply hurt by the vicious attacks of one whom he had previously considered a friend. But Owen's reputation as a scientist remained under a cloud for many decades until more recent historians set the record straight. In recent years Adrian Desmond and others have rehabilitated Owen's reputation as one of the outstanding scientists of the Victorian period. Of course, Owen did have an ideological ax to grind, but so did Huxley. Whatever his personal shortcomings, Owen was perhaps the greatest anatomist of his day, and his criticisms of Darwin were not frivolous, but serious objections that had to be addressed by Darwin's defenders.

Professional historians of science now scrupulously avoid writing history in the Whig style, though it is still often encountered in popular accounts of science. Historians now make every effort to place past scientists and theories squarely in the context of their original historical setting. They also strive to tell the story of science "warts and all," recounting not just the successes of the theories that ultimately triumphed, but also giving evenhanded accounts of those that did not. In general, the stand professional historians of science have taken against the Whig style is laudable, but any good thing can be taken too far. In fact, I think that the fear of writing Whig history has grown from a healthy caution to an unhealthy phobia in the minds of some historians. Historians now are so afraid of falling prey to what writer C. S. Lewis called "the parochialism of the present," that they shy away from making any judgments about particular episodes as promoting or inhibiting the growth of science. In consequence, the history of science has become particularly susceptible to the allure of social constructivism.

SOCIAL CONSTRUCTIVIST HISTORY

Social constructivist history of science certainly is not history done in the Whig style. On the contrary, social constructivists, as we saw in the last chapter, interpret *all* scientific episodes, the "successful" and the "unsuccessful" alike, as ultimately caused by ambient social and political circumstances, not the considerations of evidence and logic traditionally thought to drive science. Or, more accurately, they see even the canons of evidence and logic employed in the sciences not as embodying universal norms of rationality, but merely as "rules of the game" radically contingent upon historical factors. If the very standards of rationality and evidence employed in science are themselves socially constructed through and through, then science cannot be a march towards truth, but only a random walk in response to the shoves and jerks of historical forces. For the social constructivists, therefore, no episodes in the history of science can be seen as "progressive," in the sense of approaching closer to truth or providing objectively better methods for the discovery of truth.

Perhaps the best-known example of social constructivist history of science is the 1986 book *Leviathan and the Air-Pump* by Steven Shapin and Simon Schaffer. Shapin and Schaffer's book focused on an important controversy from the mid-seventeenth century, the nasty dispute between scientist Robert Boyle and irascible philosopher Thomas Hobbes. The dispute centered on the efforts of Boyle and his allies to introduce experimental methods into science. These changes required the radical reorganization of the way that science was practiced and the imposition of tight controls over who was recognized as a legitimate practitioner. Today experiment is such an integral part of scientific procedure that it is hard to imagine that science was ever practiced without experimental methods. Yet such methods had to be invented and incorporated into science. Although he was a skilled dissector, Aristotle performed no experiments in the modern sense. For Aristotle, and his followers right up into the modern era, scientific method was a matter of performing "inductions" from particular instances to first principles, and then making "deductions" from those principles back to particular kinds of things (at least this was Aristotle's stated method; his own scientific practice did not always conform to his official line). Francis Bacon famously proposed the use of experimental methods in the place of Aristotelian methods, but the closest he ever came to actually performing an experiment was an attempt to preserve a chicken by stuffing it with snow (Bacon became a martyr to science when he caught a fatal case of bronchitis after his chicken-stuffing experiment). Galileo performed brilliant experiments testing the acceleration of falling bodies by rolling balls down an inclined plane. Galileo's most famous experiment was one that never took place, the famous story about him dropping cannonballs from the Leaning Tower of Pisa. However, it was Robert Boyle's experiments with the newly invented air pump, which for the first time allowed scientists to create a fairly good vacuum, that placed experimental method front-and-center in physical science. Thus it was only in the 1660s, the time of the Restoration (the restoration of the monarchy after the English Civil War) that experiment became essential to science.

Surely, it seems, the adoption of experimental methods was an instance of scientific progress if anything was. Again, it is hard for us to imagine things any other way. Shapin and Schaffer say that when Boyle first began his promotion of experimental methods in the 1660s things were not so clear. In particular, noted philosopher Thomas Hobbes objected to the experimental method on various grounds. Among other complaints, Hobbes charged that the experimental method restricted the practice of science to a specialized elite, a privileged few who were granted access to expensive equipment that only they knew how to build and operate. With Boyle we see the beginning of the professionalization of science, that is, the recognition of a properly credentialed professional elite as the only legitimate practitioners of science. Today, of course, the certification process is quite strictly controlled. If you want to have a career in science, the Ph.D. is your union card. With the exception of astronomy and paleontology, where amateurs still often make valuable contributions, science is done by the professionals. In Hobbes's view, plain everyday observation and careful deductive reasoning were all that should be needed for good science. For Hobbes, armchair science was still very much a live option. As for Boyle's research into pneumatics, the study of the properties of air, Hobbes objected that Boyle's pump was leaky and unreliable. It had to be, Hobbes argued, since he thought he could demonstrate that a vacuum was in principle impossible.

Boyle and his allies, on the other hand, were deeply suspicious of the grand speculative systems of philosophers like Hobbes, which they saw as castles on the clouds. For Boyle, as for scientists today, the problem with philosophical systems is, to paraphrase a quip of Mark Twain's, that they deliver such a wholesale return of conjecture from a trifling investment of fact (Twain originally said this about *science,* but it fits philosophy much better). For Boyle, science must be based on high-quality facts, and only rigorously controlled experiment could provide facts with the precision and certainty necessary for good science. Uncontrolled, everyday observations are just not good enough because they are so subject to uncertainty, vagueness, and error. Further, only those with the specialized and rigorous training in scientific theory and practice have the competence to conduct experiments.

Of course, Boyle and the experimentalists won this debate. Whig historians of science therefore depicted Boyle as the hero and Hobbes as a crackpot. Shapin and Schaffer object that this is far too facile an evaluation of the debate, and they contend that at the time Boyle was not so clearly right and Hobbes was not an obvious crank. In fact, they argue that Boyle's case was just not strong enough to have carried the day on rational grounds. Why, then, did Boyle win? Shapin and Schaffer answer that Boyle won because he played the political game much better than Hobbes, and because Hobbes was swimming against the tide of history. They argue that the emergence of a new way of organizing science was an integral part of the emergence of the new social order of Restoration society.

Hobbes was already at a disadvantage when he entered the dispute. He had the reputation, not entirely undeserved, of a cantankerous and crotchety old man who was always eager to keep a dispute boiling even when he was demonstrably wrong. As a philosopher, Hobbes endorsed an uncompromising materialism and his classic political treatise *Leviathan* supported absolute authoritarianism in government. Worse, Hobbes was suspected of atheism; he certainly made a number of statements that made the orthodox nervous. Boyle, on the other hand, was fervently and famously pious, and that mattered a lot in those days. Boyle also had many connections with the rich, powerful, and influential. Who you know may not count for everything in science, but it certainly counts a lot. In the end, the well-connected Boyle just had too many resources and too much influence for Hobbes to counter.

Shapin and Schaffer generalize from their case study and assert that the methods of science are always a product of local social and political influences. The very methods of science, the methods that scientists think will lead them to truth, are actually historically contingent conventions adopted on the basis of political expedience. The consequence, Shapin and Schaffer state very plainly, is that *we ourselves* determine the content of our science, not some putative external nature: "As we come to recognize the conventional and artifactual status of our forms of knowing, we put ourselves in a position to realize that it is ourselves and not reality that is responsible for what we know" (Shapin and Schaffer, 1986, p. 344). In other words, the methods of science may change, but they do not *improve* in any sense other than perhaps having more social or political utility for the people that endorse them. Clearly, if the methods of science are really just arbitrary conventions and stipulated "rules of the game," there is no clear sense in which science progresses—certainly not towards truth. So it looks like if we go too far trying not to tell the story of science as a triumphant march of steady progress, we wind up not being able to see how it can progress at all.

Shapin and Schaffer attempt to give a *causal* account for the fact that scientists sometimes adopt a new set of methods. For their account to be adequate they have to rule out the opposing hypothesis, that scientists adopt new methods for the simple reason that they see that they *are* better ways of doing science. No, Shapin and Schaffer contend, scientists adopt new methods because they are politically and socially expedient. So, for their explanation to work, Shapin and Schaffer must argue either (a) that no scientific method really is any better than any other, or (b) that scientists cannot distinguish better from worse methods.

How could anyone argue (a) that no method is better than any other? Scientists are constantly trying to work out new methods that will permit them to better test their theories against data. A new method can be better than an old one by providing a greater quantity of data, or a greater variety of data, or data more directly relevant to the hypothesis being tested, or clearer data with more signal and less noise, or more precise data allowing for a stricter, more rigorous test. Other methodological innovations might provide more incisive mathematical or analytical tools for making better sense of data. How can anyone say that no method can be better than another in any of these regards? The only way to argue this in a comprehensive way would be to adopt a universal skepticism about method. The most famous case for such skepticism is the 1975 book *Against Method* by maverick philosopher of science Paul Feyerabend. Feyerabend appeals to the history of science to argue that no methodological prescription has ever been consistently followed in science. Every methodological rule, he claims, has been violated at one time or another by even the most famous and successful scientists. Further, he asserts that the biggest scientific discoveries could not have been made had not scientists broken the rules and thought "outside the box." Feyerabend concludes his attack on method with the ironic comment that the only rule that science had consistently followed is "anything goes."

Now Feyerabend is certainly right that no simple set of exceptionless rules has guided every scientific judgment. By analogy, no set of simple, unbreakable rules covers every possible situation where we have to make a *moral* decision. Even good rules like "do not lie," and "do not steal" admit of well-known exceptions where it would be permissible or even morally obligatory to lie or steal. Also, methodological rules change over time (if they didn't, how could they ever improve?). However, Feyerabend's methodological anarchism goes way too far. As philosopher of science W. H. Newton-Smith notes, to show that a scientific method is worthless, it is not enough to cite anecdotes about its failure on one occasion or another. You would have to show that it has failed more often than not, and this task Feyerabend did not even begin to accomplish. Further, to make any judgments about a particular rule, we must, at least for the time being, assume that other rules are reliable:

> For how do we know that a particular rule has led us to make unfortunate choices? We have no omniscient God to whisper the answers in our ears. Trapped as we are within the scientific enterprise without such a divine road to knowledge, we have no recourse but to make such judgments on the basis of other principles of comparison. Thus any historically based attack on a particular methodological rule of the sort being envisaged will presuppose the viability of other such rules. The best one can do through an historical investigation is to take up a single plank of the ship of methodology while the rest remain, for the moment at least, firmly in place. (Newton-Smith, 1981, p. 134)

Newton-Smith says that rejecting a methodological rule is like taking a plank out of a ship at sea. We can replace a bad plank, but only if we leave the other planks in place for the time being. If we try to take all the planks out at once, we have no ship left and sink to the bottom. Likewise, we can reject a particular methodological prescription only if we rely on others that, for the time being, we do not question. Trying to reject all methods at once would leave us helpless to make any judgments. Therefore, the idea that we could reasonably be skeptical about *all* methods at once is just incoherent.

So Shapin and Schaffer cannot reasonably maintain a global skepticism about all methods. Therefore, they would have to claim (b), that, though improved methods might be proposed, scientists are incapable of recognizing them as superior to other methods. Outside of postulating a universal curse laid on all human cognitive powers—in which case their *own* reasoning would be in doubt—it is hard to see how Shapin and Schaffer could argue this. Suppose that double-blind trials really are better ways of testing the effectiveness of new drugs than gazing into crystal balls. What could prevent sane, intelligent humans from learning this fact?

Perhaps, then, Shapin and Schaffer would concede that some methods are better than others, and that scientists can sometimes recognize that a particular method is better than another. However, they would argue that in real life scientists seldom adopt new methods because they see that they are better than the old ones, but because they give in to political and social pressures. In other words, perhaps their position is not so much skeptical as cynical. However, such cynical conclusions would have to be argued on a case-by-case basis. They certainly cannot justify universal cynicism about science by generalizing from a single case study—though this is precisely what they try to do in *Leviathan and the Air-Pump*.

Another problem with the cynical thesis is that there are so many apparent counterexamples from the history of science, that is, instances where scientists without political clout or social standing proposed new methods that soon were widely accepted. For instance, in 1908, Henrietta Swan Leavitt, a low-level employee at the Harvard Observatory, noticed something remarkable about a particular kind of star. She was examining stars known as Cepheid variables in the Small Magellanic Cloud, an irregular companion galaxy to the Milky Way. These stars were called variables because they go through a regular cycle of getting brighter and then dimmer and then brighter again. She noticed something peculiar, namely, that the brighter a Cepheid variable was, the longer was the period of its cycle from brightest to dimmest. Since all the stars in the Small Magellanic Cloud are about the same distance from us, the stars there that *look* brighter really *are* brighter. So, Cepheid variables can serve as "standard candles"; that is, when you see a Cepheid variable in a distant galaxy, you can measure the period of its brightness variation, and then you know how intrinsically bright the star is. Knowing how intrinsically bright a star is, and comparing that to how bright it looks, can tell us how far away the star is, and so how far away its galaxy is. So, Leavitt had discovered facts that gave astronomers an invaluable tool for measuring celestial distances. Astronomers had nothing socially or politically to gain from adopting the method based on Leavitt's discovery of the period/luminosity relationship; she was a woman when women had little status in science and she certainly had no important political connections. Clearly, the reason that astronomers soon adopted methods based on Leavitt's discovery was that those methods were *good methods.*

So, have historians of science, perhaps due to excessive fear of writing Whig history, failed to address the issue of progress in science? Noted philosopher of science Larry Laudan thinks so. He says that if a historian admits that some theories have had more empirical success than others (such as making more accurate predictions), and this is hardly deniable, then the issue of progress should be addressed:

> If he once grants that certain scientific theories or approaches have proved more successful empirically than their rivals, then the historian who disavows any interest in scientific progress is confessing there are some facts about the past which he has no interest in explaining. Given that no history can be complete, that alone would not be very distressing. But the progressiveness of science appears to everyone *except the professional historian of science* to be the single most salient fact about the diachronic development of science. That, above all else, cries out for historical analysis and explanation. (Laudan, 1990, p. 57; emphasis in original)

A historian of science who ignores the issue of progress is like a historian of the American South who ignores the issue of race. One of the, if not *the*, most important issues of the field is simply dismissed.

But how can we designate certain episodes in the history of science as more progressive than others without privileging some set of criteria about what constitutes progress, and wouldn't such judgments inevitably be Whiggish? By whose standards are we to judge that progress has or has not taken place? By *our* standards, of course. As Newton-Smith noted above, we hold, and are bound to hold, that some methods for obtaining knowledge are better than others. The only alternative is an attitude of complete indifference about all knowledge claims. Such an attitude is not in evidence among historians of science, not even—perhaps we should say "especially not"—among those inclined towards social constructivism.

So, if past scientists reached a conclusion we still think is true, or approximately so, and if they based their conclusion on reasons that still seem good to us today, we should regard this as an instance of progress in science. Is such a judgment Whiggish? The problem with Whig history is that it is unhistorical; it judges people by standards inappropriate for that time and place. But if we judge the theory-choice decisions of past scientists in terms of the reasons available *to them,* the fact that those reasons are still good for us today does not make such judgments unhistorical. It can never be unhistorical to tell the truth about history, and, as Laudan asserts, perhaps *the* salient truth about science is that it has progressed. If so, to approach the history of science as the social constructivists do, with the grim determination to explain away any appearance of progress, is as blatant an instance of ideologically blinkered ahistoricism as anything ever perpetrated by the Whigs. Laudan states this point eloquently, referring back to Herbert Butterfield's original definition of "Whig history":

> One can sympathize with Butterfield's concern that a tale of victory, told only by the victors, makes for bad history. But in denying that historians are ever justified in recognizing that certain parts of science are better than others, and in asserting that it is no part of the historian's task to explain the conditions which made them more successful, Butterfield (and those historians of science who follow him) would appear to be abandoning the programme of telling the full story of the past to which they are otherwise so deeply committed. (Laudan, 1990, pp. 56–7)

DOES SCIENCE CONVERGE TOWARDS TRUTH?

Let us suppose then that science does progress. How does it progress? Laudan, who is so eloquent in defending the progress of science, is equally eloquent in arguing that there is little evidence that it progresses towards *truth*. Science could progress in various ways without getting closer to truth. It could progress in the purely pragmatic sense that our science today gives us greater control over nature and more advanced technology than past science. Science could progress by producing theories that did a better job of saving the appearances, by generating more and better predictions. As we saw with Ptolemy's astronomy, a false theory can often save the appearances very successfully. When Ptolemy said that Mars would be found at a particular part of the sky at a particular time, he was always very close to being exactly right. Science might also progress in the Kuhnian sense that a new theory can accommodate the anomalies that stumped the old theory. Traditionally, though, science was thought to progress towards truth. On this view, Newton gave us a truer view of the universe than Aristotle, and Einstein's was truer than Newton's. Do we have the final truth yet? No, because our two fundamental theories of physics, relativity and quantum mechanics, contradict each other, so something is wrong somewhere. Still, scientific optimists are convinced that successive theories *converge* towards truth, and hope that someday, perhaps, we will have a complete "theory of everything." Let us therefore call *convergent realism* the view that over the history of natural science successive theories have converged towards truth, that is, that later theories are better approximations of truth than earlier ones.

Laudan argues vigorously that the history of science confutes the idea of convergent realism. Why have people usually thought that science progresses towards truth, that as new theories replace old ones over time, the new theories get closer and closer to the whole truth? The reason usually given is that in any field of science the more recent theories are almost always more empirically successful than the earlier ones. A theory is more "empirically successful" than another if it makes more accurate predictions, or passes more stringent tests, or gives better explanations of puzzling phenomena, or allows more effective intervention in the natural world (as, for example, in better preventing disease). (By the way, by "prediction," philosophers of science just mean an observational consequence of a theory. A prediction in this sense does not have to forecast the future; it can "retrodict" something that has already happened, and even something already known.) Thus, Newton's theory made many more, and more accurate, predictions than the old Aristotelian scheme. It also permits us to do a lot more; calculations drawing on Newtonian laws are still used to send spacecraft to the outer planets. Einstein's theories incorporated and improved on the empirical successes of Newton's theories. To many minds, the spectacular increase in the empirical success of science over time is just incomprehensible unless science is converging towards truth. As a number of philosophers have put it, the ever-growing empirical success of science would simply have to be a stupendous miracle unless science were also growing towards truth. The tremendous technological advances of the past four centuries depend on the accuracy of theoretical predictions. How then is it that we can send rockets to distant planets, cure diseases once thought incurable, and probe the cosmos as it was billions of years ago, unless we have some inkling about what is really going on? If we don't really know more than people did 500 years ago, how can we *do* so much more?

The above argument certainly has great intuitive appeal, so how could anyone resist it? The most direct way would be to show that empirical success and truth have not been connected in the history of science. That is, if we could show that many past theories enjoyed considerable empirical success in their day, but are now rightly regarded as false, this would de-couple empirical success from truth. In this case, the increasing empirical success of successive theories will not show that science is progressing towards truth. On the contrary, the fact that so many theories have been shown wrong, though they enjoyed great success and nearly universal acceptance in their day, should lead us to conclude that the majority of our present theories are likely false. This argument has been called the "pessimistic meta-induction." Reasoning from particular instances to a general conclusion is a form of induction. The induction here generalizes from the history of scientific theories to the pessimistic conclusion that our current theories are probably false (it is called a "meta" induction because it does not generalize from particular facts but from theories which are already general statements—but this isn't really important). Laudan asserts that the fact that so many successful theories of the past have turned out false is good inductive evidence that our present ones are false also, and this is why the induction is "pessimistic."

Laudan refers to case after case of past theories that enjoyed considerable empirical success in their day, even making accurate predictions that were novel and surprising (which counts a lot in a theory's favor), yet these theories are now universally rejected as false. A good example would be the theory of phlogiston. In the late eighteenth century, chemists wanted to know exactly what happens when something burns. When something burns it seems to lose substance—matter flows out leaving only ashes. Eminent English chemist Joseph Priestly hypothesized that the "fire stuff," the volatile matter that flows out of things as they burn, is something called "phlogiston." The phlogiston hypothesis had a lot going for it. It accurately predicted that a burning candle placed in an airtight container would soon go out. The explanation was straightforward and made perfect sense: As the candle burns the air inside the container becomes saturated with phlogiston and the phlogiston still in the candle has no place to go, so it stops flowing out. Phlogiston theory could also explain the process of smelting whereby crude ores are refined into metals. These and other empirical successes made phlogiston theory a prominent and widely accepted chemical theory in its day. Nevertheless, phlogiston theory was ultimately rejected in favor of Antoine Lavoisier's hypothesis—that combustion is not a matter of something flowing out, but of something combining with the burning substance, namely oxygen. We still accept Lavoisier's account of combustion today, so phlogiston theory is regarded as false.

Laudan says that such examples can be adduced *ad nauseam*, that is, we can list indefinitely many instances of past theories that enjoyed considerable empirical success in their day but are now universally rejected.

We cannot even regard theories like phlogiston as partially true, says Laudan. Often a new theory does not completely supplant the old one, but retains some portion of the old theory or perhaps even absorbs it as an approximation or a limiting case. As we saw in Chapter Two, many claim that Newtonian theory remains as a limiting case of Einstein's theory of special relativity. At relative speeds far below the speed of light, Einstein's laws of motion reduce, very nearly, to Newton's. Often also a theory can be considered partially true because its central terms refer to objects still recognized as real. For instance, when Gregor Mendel first published his theory of inheritance in 1866, he postulated the existence of certain "factors" or "elements" as units of heredity. Though the science of genetics has progressed enormously since Mendel, transformed fundamentally by molecular biology, Mendel's "factors" or "elements" are still thought to refer to what we call "genes" today. Yet with theories like phlogiston, *nothing* is retained in later theories; they are rejected lock, stock, and barrel. "Phlogiston" does not refer to anything still considered real any more than does "witch." Therefore nobody can say that phlogiston theory was partially true and therefore a step up the ladder of convergence towards the true theory. The history of science is a graveyard of theories that were mighty in their day but are now nothing but dust occasionally shaken from old books.

Apparently, then, a theory can enjoy considerable empirical success even if it is not even partially true. Actually we can see this rather easily without having to make reference to arcana from the history books like phlogiston theory. The earth is not flat, but if you want to build a house, or even a large building, the hypothesis that the ground underneath the building is flat works perfectly well. The reason why, of course, is that the earth is so big that its curvature, even over the area of a large building, is negligible and the space can be presumed flat. The sun, planets, and stars do not revolve around the earth, but for purposes of navigation, a geocentric theory works as well as a heliocentric one. So, false theories can have true consequences, and this shouldn't really surprise anyone. Any student taking an introductory logic course soon realizes that false premises can entail true conclusions. From the false premises "All Democrats are Republicans" and "Ronald Reagan was a Democrat," it follows as a valid deductive consequence that "Ronald Reagan was a Republican," which is true.

Defenders of convergent realism are, of course, aware that false theories can have true consequences, and therefore that false theories can be expected to have some empirical successes. Their argument is that the truer a theory is, the *more likely* it is to have true consequences, and so, the *best explanation* of the increasing empirical success of science over time is that successive theories are better and better approximations of truth. Further, convergent realists argue that new theories typically do retain and improve upon elements of the theories they replace. They conclude that the best explanation for the empirical success of earlier theories is that the parts of those theories retained in later theories express at least partial truth about physical reality. Einstein did not retain all of Newton's ideas. For instance, Newton claimed that space and time are absolute, which Einstein explicitly and emphatically rejected. Convergent realists think that the parts of Newtonian theory that Einstein did retain as a limiting case are the parts that give us a very good approximation of the truth about the motion of bodies at relative velocities far lower than the speed of light. They also think that it is those parts that were retained, not the ones Einstein discarded, that explain the great empirical success of Newton's theories.

Laudan replies that the connection between partial truth and empirical success has not been established. True statements cannot entail a falsehood, but Laudan thinks that no one has shown that, in general, a theory 50 percent true (whatever that means) is more likely to have true consequences than one 0 percent true. Further, he denies that as a rule new theories retain significant elements of the theories they replace. So, Laudan challenges convergent realists to prove the connection between partial truth and empirical success and to show that the history of science supports their factual claims about successive theories.

ASSESSING LAUDAN'S CRITIQUE OF CONVERGENT REALISM

Whether or not partially true theories are more likely to enjoy empirical success than wholly false ones, it is clear that the history of science has *not* been a tale of steady ascent towards truth. Such a tale is what we earlier called Whig history, and we have already seen that such an interpretation of the history of science is distorted and inaccurate. Science is not a straight-line ascent to the peak of Mount Truth. Even those who think that science has acquired some deep truths about the cosmos must view the history of science as a difficult scramble up a slippery and treacherous slope, with many halts, detours, slips, and stumbles along the way. So, if convergent realism requires a steady march of scientific progress towards truth, then the history of science refutes convergent realism, just as Laudan claims. But is it possible to take a view more consistent with the actual history of science and still see science as progressing towards truth? How can we generalize about the history of science without running into the sorts of rather obvious counterexamples Laudan adduces—theories very successful in their day but now recognized as wholly false?

As Laudan interprets the case for convergent realism, it claims that the best, perhaps the only, explanation for the empirical success of science is that current theories (at least in the "mature" sciences like chemistry or physics) are typically approximately true, and more recent theories are truer than their predecessors. We can interpret Laudan's critique as making a strong claim and a weaker claim. The strong claim is the pessimistic meta-induction: that even our best-supported current theories should not be regarded as approximately true given the historical record of highly successful theories that were wholly false. If, typically, even the most successful past theories turned out completely false, then most of our currently accepted theories are probably not even approximately true, despite their enormous empirical success. Laudan's weaker claim is that convergent realists cannot argue that the increasing empirical success of theories over time supports their assertion that science converges towards truth. He contends that the argument fails because convergent realists have not established the necessary correlation between degrees of truth and degrees of empirical success. Laudan complains that we have no philosophical reasons to expect such a correlation, and he contends that the facts of the history of science are against it.

Starting with the stronger claim, we first have to be clear on what convergent realists are claiming. They are happy to concede that, strictly speaking, many of our best current theories are false. Convergent realists claim that these theories, though false, are nevertheless approximately true. For instance, convergent realists can regard the currently accepted standard model of particle physics as false, strictly speaking, but still as approximately true. Critics such as Laudan often complain that no one has yet articulated a philosophically satisfactory notion of approximate truth. Actually, since philosophers have yet to agree on the meaning of "truth" it is hardly surprising that there is no consensus about the meaning of "approximate truth." Still, we have very deep and undeniable intuitions that some claims are truer than others, or, in the words of Isaac Asimov, that wrong is relative. It is wrong that the earth is a sphere; actually, it is an oblate spheroid that bulges slightly in the middle and is slightly flattened at the poles (from space it still looks like a perfect sphere, though). But is it just as wrong to say that the earth is a sphere as to say it is a cube? Clearly, we have a very strong intuitive sense that the statement, "the world is a sphere," while false strictly speaking, is more "truth-like" than the statement "the world is a cube."

Philosopher Stathis Psillos defends a notion of approximate truth that builds on these intuitions about truth-likeness (1999). Psillos notes that in science theories are seldom even presented as *exactly* true (more on this in the next chapter). Rather, scientific theories generally offer representations of the world that are idealizations, simplifications, or approximations rather than assertions purported to be precisely true. Even statements of the laws of nature contain clauses that bracket off and exclude distorting factors, like the effects of air resistance on free fall. Yet a theoretical description can still be highly truth-like, even if not strictly true, when it *fits* reality to a high degree of approximation. Psillos spells out these intuitive ideas more precisely as follows:

> A description D *approximately fits* a state S (i.e., D is approximately true of S) if there is another state S′ such that S and S′ are linked by specific conditions of approximation, and D fits S′ (D is true of S′). (Psillos, 1999, p. 277)

In other words, a truth-like theoretical description will be one that is *strictly* true of an imaginary state, but that imaginary state is one that is very similar to the relevant state of the actual world.

Consider the imaginary world in which Newton's laws of motion hold *precisely* in all circumstances; let's call this imaginary world "Newton-World." Newton-World is not the real world. We now know that Newton's laws fail in various circumstances, and physicists have given us relativity theory and quantum mechanics to deal with the realms where Newton's laws fail. Yet Newton's laws remain indispensable tools for applied physics. For instance, space scientists employ Newtonian calculations to plot the courses of spacecraft through the solar system. When you send a probe to Neptune and it arrives within seconds of the predicted time, you are doing *something* right. So it looks like Newton-World is, in many important respects and under many different conditions, very much like the real world, and Newton's theory, though false, still has a high degree of truth-likeness.

Now Psillos admits that such an intuitive formulation of the nature of truth-likeness will not satisfy philosophers who want a formal and rigorous statement of truth-likeness. Yet he does not regard the lack of such a rigorous formal statement as an insuperable problem for realists. I agree with him. If we had to wait until philosophers formulated rigorous accounts of all of our key concepts, we would not get much done. So, I shall assume that the notions of truth-likeness and approximate truth are clear enough for our discussion to proceed.

Getting back to Laudan's pessimistic meta-induction, all such inductions are based upon samples. Such arguments conclude that what is true of a sample is, to a high degree of approximation, true of the whole from which the sample is drawn. For instance, Laudan offers instances of successful theories that turned out wholly false and bases conclusions about the whole history of science on his sample. To reach a true conclusion, inductive arguments of this sort must draw upon samples that are truly representative of the whole. Inductive arguments can go disastrously wrong if proper steps are not taken to assure that the sample is representative of the whole. Defenders of the pessimistic meta-induction therefore need to show that the *typical* fate of empirically successful theories of the past has been to be proven *totally* false, i.e., not even partly true. Only if they show this can they effectively argue that today's successful theories are likely to be wrong. Laudan has not shown this. Even if we can adduce examples of successful totally false theories *ad nauseam,* as Laudan claims, this does not prove that this list is representative of all past theories. We can also list *ad nauseam* instances of past theories that, while woefully wrong on some things, got other things nearly right, or at least referred to entities science still regards as real.

However, even if we for the moment concede, purely for the sake of argument, that the majority of successful past theories were totally false, the pessimistic meta-induction is not necessarily warranted. The reason is that the quality and quantity of the empirical tests theories must pass has greatly increased over the history of science. One way that science unquestionably progresses is that new methods and techniques are frequently found that allow for ever more stringent tests of theories. Judged by the kinds of empirical tests and analytic tools we have today, many past theories were not very well tested, and so were not really very successful compared to current theories. On the other hand, the Santa Claus hypothesis is highly empirically successful for five-year-olds, but, of course, five-year-olds are very limited in their ability to test hypotheses. The point is that a theory's degree of empirical success is highly dependent upon the severity of the tests it must pass and the stringency of the standards it must meet.

Even in a "soft" science like paleontology, the number and the rigor of empirical challenges that theories must meet are far greater than in the past. For instance, the idea that dinosaurs might have been warm-blooded, rather than cold-blooded like typical reptiles, was first proposed by T. H. Huxley in the mid-nineteenth century. At the time, Huxley's proposal was not much more than a speculation since data bearing on dinosaur physiology were sparse and very indirectly relevant. In the 1970s Robert Bakker revived the idea of warm-blooded dinosaurs, and vigorously prosecuted his case. By then the quantity and quality of evidence that could be brought to bear, and the sophistication of methods and techniques for judging Bakker's claim, vastly exceeded anything available to Huxley. Bakker employed microscopic studies of bone histology to argue that dinosaurs had bone architecture closer to that of warm-blooded mammals and birds than to cold-blooded reptiles. He appealed to predator/prey ratios in fossil assemblages to argue that the proportion of predators to prey among dinosaurs was like the ratio of warm-blooded lions to prey in the Serengeti, not the ratio of cold-blooded komodo dragons to their prey. More recently, tests on the ratios of oxygen isotopes in dinosaur bone have been used to test the hypothesis of warm-blooded dinosaurs. Needless to say, Huxley could not even have dreamed of having such evidence. Even whole new sub-disciplines—like taphonomy, the study of the process of fossilization, and paleoichnology, the study of fossil tracks and traces—have been developed to provide more tools for the evaluation of theories in paleontology.

The upshot is that the success of false past theories is due largely to the fact that they were not tested nearly as severely as we test theories today. Hence, the generalization the pessimistic meta-induction rests on, the claim that many, perhaps most, successful past theories turned out to be wholly false, is not strong enough to show that our currently accepted theories are not approximately true. The "success" of many of those past theories is just not comparable to the success of today's accepted theories. Laudan, for all his emphasis on scientific progress, seems not to appreciate the degree to which, in any given field of science, the sophistication and rigor of hypothesis-testing practices can greatly increase in a relatively short time. Does this mean that we have crossed some sort of threshold and that our tests are now so rigorous that we can be *sure* that we will keep any totally false theory from slipping by? Of course not, but convergent realists can say with considerable confidence that a totally false theory would have a much tougher time hiding its false consequences today than two hundred years ago.

What convergent realists would therefore expect to see when they look at the history of science is that wholly false theories would proliferate when empirical tests are weak, but would enjoy less and less success as tests get progressively stronger. Instead, they would expect that when more recent theories are overturned, after reigning for a time in their fields, they would not so often be totally rejected, but would more often be partially preserved in the new theories. Does the history of science conform to these expectations? This is far too big a question to be settled here, but I think we can say that *prima facie* it does, or at least that the burden of proof is on Laudan to show that it does not.

What about Laudan's weaker claim that the history of science fails to support the convergent realists' argument that the best explanation of the increasing empirical success of science is that it converges towards truth? If "convergence" is interpreted in the Whig way, Laudan's argument is airtight. The history of science has not been the steady accumulation of ever-closer approximations of theories now regarded as nearly true. Rather, in each field of science there were long periods dominated by empirically successful theories (successful by the standards of their day) but now wholly rejected. But could a realist, one who thinks that at least some of our best current theories are approximately true, still find support from the history of science? What would a more realistic realism, one not mired in Whig fantasies about steady progress, expect to see in the history of science?

The history of science is not one of steady cumulative progress, but neither is it a succession of mutually exclusive paradigms where each new theory totally wipes the slate clean and starts all over again. If we regarded all past theories as totally false, then the pessimistic meta-induction probably should make us doubt our present theories, however empirically successful they are. But the history of science is not always like the famous Peter Arno cartoon from *The New Yorker:* A test flight has just ended in a horrendous crash. The aircraft designer turns his back on the ensuing chaos, claps his hands together, and blithely chirps, "Well, back to the old drawing board." Science does not have to go back to the drawing board with every superseded theory. Rather, when we look at the history of any field of science, a few theories will stand out as major breakthroughs. Once these breakthroughs occur, they are retained, in one form or another, through all subsequent theory changes, even through major conceptual revolutions. For instance, the mathematician and physicist James Clerk Maxwell (1831–1879) formulated a small set of simple equations that explained all the diverse phenomena of electricity and magnetism. He concluded that electricity and magnetism were different aspects of the same force, electromagnetism, and that light is actually a form of electromagnetic radiation. Maxwell's *Treatise on Electricity and Magnetism* was published in 1873, well before the two major revolutions in twentieth-century physics, relativity and quantum mechanics.

The revolutions of twentieth-century physics overthrew some of Maxwell's ideas. For instance, he thought that since light was a wave, it had to be carried by some medium, the "luminiferous ether," an idea rejected by subsequent theory. However, light is still regarded as electromagnetic radiation, and Maxwell's equations, in modified form, are still regarded as valid for a given range of electrical and magnetic phenomena. Likewise, Newton's famous law of universal gravitation is retained in current physics as correctly applying to things not moving too fast and to gravitational forces that are not too strong. So, many of Maxwell's ideas, like Newton's, have survived the enormous conceptual upheavals of the relativity and quantum revolutions, revolutions that overthrew so many of the ideas of "classical" physics. Within limited contexts, Maxwell's and Newton's theories are just as valid as they ever were. Other breakthrough theories have shown similar staying power in other fields of science. Lavoisier's theory that combustion is an oxidation reaction is still accepted despite the revolutions in chemistry and physics since the eighteenth century. Darwin's theory of natural selection and Mendel's genetics have been absorbed, considerably modified and augmented, into current biological theory. The retention of key elements of past theories into the present does not always happen; it may not even be the typical case. But the fact that it has happened as often as it has—that so much has been retained so often, even across scientific revolutions and through paradigm shifts—is a remarkable fact about the history of science. What accounts for it?

Realists have a simple explanation for the remarkable persistence of certain theoretical ideas: Newton, Darwin, Lavoisier, Mendel, *et al.,* were on the right track. They did not have the whole truth, but they had some of it. Human thought has enormous centrifugal force. Ideas tend to fragment and to disperse into mutually exclusive and usually mutually hostile schools, dogmas, ideologies, and "isms." We see this often in the history of philosophy, where each major philosopher saw his system as the one true one, displacing all earlier theories. Religious doctrines are notoriously brittle, easily fragmented by schism and heresy. This is why religious bodies have so often employed heavy-handed tactics like inquisitions and crusades in attempts to impose unity. In science, to a greater degree than in any other human intellectual endeavor, we find a centripetal force that is strong enough to overcome the centrifugal pull of clashing ideas. Time and time again, scientists are pulled back to certain ideas. Indeed, they are often hauled back kicking and screaming.

Consider Friedrich Ostwald, a prominent German chemist of the nineteenth century. Ostwald held that scientists should believe only in what they could directly measure, and therefore should regard undetectable theoretical entities such as atoms as useful fictions. In 1905, the same year (and in the same issue of the same journal) that he published his first paper on relativity, Einstein also published a paper explaining the puzzling phenomenon of Brownian motion. Tiny particles of dense materials do not sink to the bottom when placed in water, but remain suspended and jiggle about in a random manner; this random jiggling is called Brownian motion. Einstein assumed that the Brownian motion of the particles was caused by the collision of the particles with innumerable, invisible, and randomly moving water molecules. His equations, based on that assumption, gave precise predictions about the expected distributions of tiny particles when suspended in water. A few years later, French physicist Jean Perrin performed careful observations of the distributions of tiny particles suspended in a drop of water and found that these precisely matched Einstein's predictions. Not only that, but Perrin was able to calculate how big water molecules would have to be to produce the observed effect. When confronted by such startling evidence Ostwald threw in the towel and admitted the reality of atoms.

Although it might surprise us now, because atoms are so fundamental to our view of the world, Ostwald was hardly the only anti-atomist in the history of science. A very fat book could be (and probably has been) written on the history of anti-atomism. The idea that matter is ultimately composed of tiny discrete bits was abhorrent to many thinkers, who opposed atomism with might and main. Religious thinkers have often associated atomism with atheism, so atomism often had to face both scientific and religious opposition. The two founding figures of Western philosophy, Plato and Aristotle, were ardent anti-atomists. But the idea just would not go away. It survived every attempt to discredit it, even reemerging, as Laudan notes, after long periods when opposing theories enjoyed greater empirical success. Needless to say, in the 2,400 years since the idea of atoms was first proposed by the Greek philosophers Leucippus and Democritus, the concept of the atom has been radically altered, even losing its original defining feature of being impossible to split ("a/tom" means "uncuttable" in Greek). Yet the idea has persisted through all of the revolutions in science over the last 2,400 years.

For realists, the only reasonable explanation of the peculiar persistence of the atomic idea, its triumph through revolutions and in spite of the efforts of some of humanity's best minds to discredit it, is that there is something right about it. There must really be atoms, and we must really know something about them. It looks like we are confronting a fundamental fact about nature, a fact that no human ingenuity or ideological rancor can eradicate: Things are made of atoms.

Realists can therefore say that we find in the history of science just what we expect to find if humans have in fact uncovered some deep truths about the hidden nature of the universe. Realism does not predict that the history of science will have developed in the Whig way. Realists need not expect that science would quickly get on the road to truth and never deviate thereafter. Rather, historically sensitive realists would expect to see that some ideas have survived in spite of everything; i.e., that some ideas time and time again have buried their would-be undertakers. Further, they would expect that these oddly persistent ideas, though perhaps eclipsed for decades or even centuries by opposing ideas, would eventually emerge stronger than ever. They would expect that successive conceptual revolutions, though changing many concepts in fundamental ways, would still have to accommodate those persistent ideas. We do seem to find all these things in the history of science, and realism seems to give a straightforward explanation of why we do.

Laudan's response is that, even if he does concede that realism accounts for certain salient features of the history of science, as a point against antirealism, this argument begs the question. Laudan notes that antirealists have always contended that the realists' argument from a theory's empirical success to its approximate truth rests upon a simple fallacy, the fallacy of affirming the consequent. Every introductory logic text tells us that it is fallacious to argue like this:

> If P, then Q.
> Q.
> Therefore, P.

For instance, this is not a valid argument:

> If it rained all night, then the streets are wet.
> The streets are wet.
> Therefore, it rained all night.

The conclusion does not follow. The streets could be wet for some other reason; maybe a water main broke. Laudan says that realists argue in the same fallacious way:

> If theory T is approximately true, then it will be empirically
>   successful.
> Theory T is empirically successful.
> Therefore, Theory T is approximately true.

When asked to justify their realism, given that it seems to rest on the fallacy of affirming the consequent, Laudan says that realists respond with just another instance of that same fallacy:

> If realism is true, then it will account for the main features of the
>   history of science.
> Realism does account for the main features of the history of science.
> Therefore, realism is true.

However, says Laudan, if antirealists do not think that a scientific theory is shown approximately true by its true consequences, they surely will not accept that a philosophical theory—realism—is true because *it* has true consequences! To expect otherwise is blatantly to beg the question against antirealists.

SCIENTISTS' OWN REALISM

Do realists argue fallaciously when they defend the approximate truth of past or current theories? Surely at this point some perceptive readers will be asking a deeper question: Do scientists really need the help of *philosophers* in articulating grounds for regarding their theories as (approximately) true? To many, the suggestion that philosophers could offer such help is like saying that philosophers could advise Lance Armstrong about cycling. To suggest that scientific claims, about atoms for instance, need an additional defense over and above the sorts of evidence that Perrin and others have given seems to me already to concede half the case to antirealists. My own view is that philosophers have very little to offer in the way of *positive* arguments for the approximate truth of past or currently accepted theories. Arguments from the history of science, however sound, cannot add significantly to the reasons scientists have *right now* for believing in atoms. No such arguments will be as strong as Perrin's evidence that convinced the anti-atomic skeptic Ostwald. I think the chief role for realist philosophers is to rebut the skeptical arguments of antirealists, arguments like the pessimistic meta-induction. In short, the best reasons for regarding theories as approximately true are not the meta-scientific arguments of realist philosophers, but the reasons scientists themselves give.

Why then do *scientists* sometimes come to regard particular theories as approximately true? The first thing to note is that scientists do not generally argue in the fallacious manner Laudan attributes to philosophical realists. They do not proceed by taking an isolated theory T, drawing observational consequences from T, and then concluding that T is approximately true since its observational consequences are actually observed. A theory must have such empirical success before scientists will accept it, i.e., the observational consequences we draw from the theory must be consistent with the actual data. But that is not enough. It is a slogan among some philosophers that "theory is underdetermined by data." What this means is that for any batch of data we might cite, infinitely many possible theories could make predictions consistent with those data. In the abstract realm of sheer logical possibility, where everything not self-contradictory is possible, this thesis of underdetermination is correct. If all that matters is that a theory's predictions be compatible with the known data, then infinitely many theories are possible. Another way of expressing underdetermination is this: For any theory that could be the correct explanation for some batch of data, it will always be possible to "cook up" an incompatible theory that also accounts for the data. For instance, any batch of natural phenomena could be explained as the handiwork of a divine creator, or two divine creators, or three creators, or 137,890,546 creators . . . and so forth.

However, scientists do not find that their challenge is to select from among an infinite number of theories that would account for their data. On the contrary, their biggest problem all too often is coming up with even *one* theory that will do the job! The reason is that it is very hard for a conceivable theory even to qualify as a candidate for truth in science. True, a new theory must promise empirical success. For instance, when Wegener first proposed continental drift, he showed that it offered cogent explanations for the "fit" of continents, the distribution of fossils, and geomagnetic data. But again, this is not enough for scientists even to begin to take a possible theory seriously. A candidate theory must be logical, coherent, and consistent with the accepted (or at least rationally *acceptable*) standards and values of scientific theorizing. A "cooked up" theory will not even make it to the list of candidates. For instance, a theory cannot drag in a lot of arbitrary (what philosophers call "ad hoc") supporting hypotheses with it, nor should it "explain" things in scientifically unacceptable ways—by appealing to magic, say. Candidate theories must also be consistent with the large body of accepted background knowledge that is presumed true by all qualified practitioners of a field. For instance, a new theory in paleontology that ignored continental drift would not be taken seriously, since paleontologists now presume that drift occurred. So strict are these requirements that at any given time in any field of science there will be a fairly small set of theories that are considered the only viable candidates for explaining some batch of phenomena. That is, scientists are confident that the (approximate) truth lies somewhere within the given range of candidate theories.

Once scientists have narrowed the field to a tractable set of candidate theories, the theories are tested against one another. If, in the end, one theory emerges above all the others, scientists then accept that theory as the right one—for the time being and until a better theory comes along. The reasoning here is not fallacious at all. Scientists assemble a small set of candidate theories, which they hold to be the only reasonable candidates, and eliminate all but one. The logical skeleton of this way of arguing is this:

T_1 or T_2 or T_3 (suppose these are the only candidate theories).
Not T_1.
Not T_2.
Therefore, T_3.

This is a perfectly valid form of argument. So, it is not just because a theory is empirically successful that scientists (tentatively) regard it as approximately true, but because it is the one that is most successful compared to a field of the only other acceptable candidate theories.

Typically, therefore, theories in science are tested against other theories, and not evaluated on their own merits alone. Let's consider an example illustrating this process of theory selection from a small set of competing candidates. Recall the debates over dinosaur extinction examined in Chapter Two. Until 1980, the most respectable theories of mass extinction postulated a series of gradually occurring processes that led eventually to the severe deterioration of the dinosaurs' environment. According to these theories the late Cretaceous climate slowly became much drier and cooler as shallow seas receded, turning the dinosaurs' lush subtropical environments into savanna or semi-desert. Smarter, faster, warm-blooded, and far more prolific mammals preyed on dinosaur eggs and could thrive in environments hostile to dinosaurs. Freezing winters and smart mammals eventually proved too much for the cold-blooded, dim-witted dinosaurs. In the extinction debates of the 1980s and 1990s, two neo-catastrophist theories rose to challenge the old uniformitarian paradigm, the impact and volcanic theories. For many scientists during this period it became clear that whatever had happened to the dinosaurs had occurred at least relatively quickly, so the old gradualist theories could not be right. Further, there was evidence that some sort of cataclysm *had* gripped the earth at the end of the Cretaceous. For instance, at many sites an abundance of "shocked" quartz grains were found in the rocks marking the Cretaceous boundary (and not in adjacent rock). Only quartz crystals subjected to extreme percussive force exhibit the characteristic deformations that lead geologists to call them "shocked." Impact theorists naturally saw this as evidence for the impact of a massive extraterrestrial body. Other scientists argued that shocked quartz is produced by ultra-violent volcanic explosions, explosions even greater than the ones that destroyed Krakatoa in 1883. No other known natural forces can produce shocked quartz.

The upshot is that by about 1990, there were three, and only three, types of extinction theory vying for supremacy: the older gradualist view (seemingly on its way out), and the neo-catastrophist impact and volcanic theories. One complication is that there were different specific theories within these broader types of theory. For instance, David Raup's "Nemesis" hypothesis was one of the impact theories. Another complication is that some scientists opted for theories combining elements of the different theory types. Thus, some paleontologists said that dinosaurs were stressed by a gradually deteriorating climate and that the final cataclysm, whether volcanic or extraterrestrial, delivered the *coup de grace*. Nevertheless, by around 1990 there was a fairly small and reasonably well-defined set of theories accounting for the end-Cretaceous mass extinctions. Why just these theories? After all, innumerable other theories could account for the dinosaurs' extinction. Maybe the dinosaurs were hunted to extinction by extraterrestrial big game hunters. Perhaps dinosaurs were on the ark with Noah, but just could not adapt to the post-flood environment. Why didn't these scenarios get equal billing with the other theories? Well, these two are just too silly for serious scientific consideration. The fact is, though, that there are severe empirical constraints limiting the kinds of plausible extinction theories that can be offered. Mass extinction is hard to achieve. Only a limited number of known natural occurrences could do it and not sterilize the earth completely, and only a few of those are consistent with the facts admitted on all sides. As we said before, even coming up with a good *candidate* for an acceptable theory is a hard job.

So, scientists tentatively accept a theory as approximately true only if (a) it is empirically successful in an absolute sense (it explains the facts it is supposed to, is supported by background knowledge, passes rigorous tests, etc.), and (b) it is *more* empirically successful than every other plausible rival they can think of. "Aha!" antirealist philosophers will exclaim, "What makes scientists so sure that they have thought of every possible acceptable rival to the winning theory? How do they know that the true theory is not one they never even thought of?" The answer, of course, is that they don't know for sure that they have thought of every possibility (more on this in the next chapter). This is not a trivial problem. Nature is full of surprises, and the experts have been left with red faces on many occasions—even Einstein. When he formulated his general theory of relativity, Einstein was chagrined to find that it implied that the universe is expanding. Einstein was so sure that the universe is not expanding that he introduced a "cosmological constant," a fudge factor, into the theory to offset such expansion. Within a couple of decades Edwin Hubble had discovered that the universe is indeed expanding. There is a well-known photograph of Einstein peering pensively through Hubble's telescope while Hubble smugly puffs his pipe in the background. You can almost see the thought balloon over Einstein's head reading "Well, I'll be damned!" Human mental powers, even the collective powers of the best and the brightest, are anything but foolproof.

Still, there are times when we can feel fairly confident that we have considered every reasonable possibility. How? The answer lies in the cumulative nature of scientific knowledge. Scientists take it for granted that we already have some knowledge about how the universe works. We have some acquaintance with natural entities, forces, and processes, and a pretty good idea of the kinds of effects they have. Scientists think that we have grasped some of the laws of nature. Getting back to the example of dinosaur extinction, I think that scientific thinking on this topic goes something like this: Something killed off the dinosaurs. Despite the fondest wishes of many a seven-year-old, they are no more. Was the extinction gradual or sudden? If sudden, as the evidence now indicates, was it caused by something on the earth or something extraterrestrial? What earthly forces could possibly produce a cataclysm sufficiently intense to account for the end-Cretaceous extinctions? Massive volcanism seems the only likely candidate; we know enough about volcanoes and their effects to realize the enormous environmental change they can cause. No other earthly force we know of could do the job and not clash with the known facts. We also know that enormous volcanic eruptions did occur at the end of the Cretaceous. If the extinction was caused by something extraterrestrial, what could it be? By far the most likely candidate, given what we know, is a comet or asteroid striking the earth. We know the laws of physics that tell us the force such a strike would generate, and we can calculate the damage it would cause. Further, other than such an impact, no other cataclysmic cosmic event that we can think of, like a nearby supernova, could possibly cause mass extinctions and be consistent with the evidence. Therefore, the end-Cretaceous extinctions were caused either by gradual deterioration of the environment, extremely violent volcanism, or the catastrophic impact of an object from space (or some combination of these factors).

Of course, we *could* be completely off base and the real reason for the mass extinctions could be something we never thought of. This is why scientific knowledge is always tentative and scientific claims always deserve some (sometimes very small) degree of skepticism. However, earth scientists would be much more skeptical of the suggestion that there might have been an unknown natural force sufficiently powerful to cause mass extinction but which acted in a way such that its effects are indistinguishable from the effects of known causes (just what did produce those shocked quartz grains?). At the very least, they will demand a plausible suggestion about what that unknown extinction agency might be before they take such a "possibility" seriously. In the meantime, they will feel justified in thinking that the truth lies somewhere in the neighborhood of our current theories.

COULD WE BE WRONG ABOUT EVERYTHING?

Hard-bitten antirealist philosophers are, of course, unimpressed by scientists' confidence in what they think they know. They doubt that scientists have grounds for claiming the approximate truth of the background theories that scientists invoke to narrow down their current candidate theories. For philosophers, unlike scientists, absolute skepticism is a genuine fear. Ever since Descartes, at least, philosophy has been haunted by the idea that, in the words of the classic Firesign Theater album, "Everything you know is wrong!" (Let me hasten to add that, since if we *know* something it must be true, it is incorrect ever to say that what we *know* is wrong—only what we *think* we know, but Firesign Theater's way of saying it is a lot funnier.) Maybe the whole edifice of science is wrong and conceivably an entirely different set of theories could be produced that would be as empirically successful as our current science. Perhaps we could encounter a race of extraterrestrials whose science was even more successful than ours, yet their theories would be incompatibly different from ours.

Still, let us pause to consider what a breathtaking claim it is to suggest that, not just a particular theory, but perhaps the whole fabric of scientific theory could be wrong. What makes this a breathtaking claim is not just the enormous empirical success of individual theories, but the fact that diverse theories fit together so well in a mutually supporting network to give us a highly coherent picture of the world. In every field of intellectual inquiry, even the most mundane, *coherence* is taken as an indicator of the probable truth of a set of beliefs. A highly coherent account will be one where the different pieces of the story all fit together so that each bit reinforces the other to create a compelling overall picture. Readers of mystery stories are familiar with how this works. At first the Inspector entertains a number of different hypotheses about who the murderer could be. Eventually, though, all the pieces of evidence and conjecture come together to constitute a compelling case against one suspect. Similarly in science different theories often reinforce each other so that each becomes far more credible than it would have been alone. We shall see below that this was the case with Darwinism and Mendelian genetics, which initially seemed opposed but were later shown to be mutually supporting.

W. V. O. Quine, one of the leading American philosophers of the twentieth century, liked to say that hypotheses face the tribunal of experience not alone but in batches. What he meant was that beliefs, whether in science or ordinary life, form an intricately interconnected network so that evidence that supports or undermines one belief also impinges upon many other connected beliefs. Thus, when evidence leads us to give up one theory, we also have to rethink other theories and make modifications or adjustments where necessary. For instance, the Cepheid variable stars, which, as we noted earlier in this chapter, Henrietta Swan Leavitt found to be "standard candles," are not bright enough to help astronomers estimate truly vast cosmological distances. Astronomers think that type 1a supernovas are also standard candles, that is, they all have about the same brightness. Type 1a supernovas are so bright that, for a time, they outshine all the other stars in their galaxy. That's why they can be seen across cosmological distances. Based on the assumption that type 1a supernovas are standard candles, astronomers have made estimates of the size and age of the universe. Also, because type 1a supernovas in distant galaxies appear dimmer than expected, astrophysicists have just in the past few years postulated the existence of "dark energy," a mysterious force that exerts a repulsive effect contrary to the pull of gravity. Clearly, then, if we came to question our theories about type 1a supernovas, we would also have to question our theories about the age and size of the cosmos since these estimates depend in part on our understanding of these supernovas. Likewise, much of the discussion about "dark energy" would have to be reexamined.

Our theories therefore do form an intricately interconnected web. Evidence that challenges one theory also challenges connected theories. It works the other way too. Evidence that supports one theory will indirectly support others. For instance, further evidence that our theories about type 1a supernovas are correct will strengthen the other theories that depend on our beliefs about such supernovas. The point is that the evidence for theories is *holistic*. The support for a theory comes not just from the empirical success of that one theory, but also from the whole network of theories and evidence to which it connects.

Philosopher Susan Haack suggests another helpful metaphor for understanding how scientific theories can mutually support one another. She compares the quest for scientific knowledge to solving a crossword puzzle. Nature provides the clues, and theories are the answers we write in. But our crossword puzzle answers are constrained not only by the clues, but also by what we have already written in. For instance, suppose that clue 7 across of the daily crossword is "literary land of little people" and the answer is eight letters. This could either be "The Shire," home of Tolkien's hobbits, or it could be "Lilliput," land of Swift's Lilliputians. The clue, by itself, does not allow us to decide between these two; the clue underdetermines the answer. Suppose though that an answer we have already written in gives us an "r" for letter seven. The answer now seems to be "The Shire." Of course the answer could still be "Lilliput," and the previous answer that gave us the "r" at letter seven is wrong. We won't be sure until we have filled in more answers and seen that they are mutually supporting.

What the crossword analogy shows is that our confidence in a particular answer (theory) depends not just on how well it fits the clue (the data) but on how well it fits in with the answers already written in (background theory). In fact, often we become really confident in an answer only after we have filled in lots more answers, and see that they all fit together. In other words, it is not only the fact that our theories are true to the data that justifies our confidence in them, but that they cohere so well in a mutually supporting relationship with other theories that we also accept. Maybe an adamant skeptic could argue that it is possible that an alternative science could have theories just as empirically successful, and just as mutually supportive. However, to take the suggestion of a complete, alternative science seriously, we have to be given more than the bare suggestion that it is possible. The skeptic would at least have to start providing us with the nuts and bolts of such an alternative science. Needless to say, no skeptic has yet achieved this. In other words, given the enormous empirical success of our best theories, *And* that they cohere in a mutually supporting network, surely the burden of proof is on the skeptic.

Looking at the history of science we do find a net increase in the empirical success of our theories, and unquestionably also an increase in the degree to which they constitute a unified, integrated, and mutually supporting network. Many of the most successful theories enjoy their status because they integrated fruitfully with other highly successful theories. Consider the historical relationship between Darwinism and Mendelian genetics. At first, there was no relationship. Darwin was already famous when *The Origin* was published in 1859; *The Origin* made him a scientific superstar. Gregor Mendel was an unknown monk who published his results in an obscure journal. Darwin apparently never heard of Mendel's results and proceeded to develop his own, completely wrong, theory of genetics. Mendel's work remained unknown until 1900 when three geneticists independently rediscovered it. It quickly rose to prominence, and, because it seemed inconsistent with Darwinian natural selection, soon sent Darwinism into a decline. It is not that scientists stopped believing in evolution *per se,* but many no longer accepted the Darwinian explanation of evolution in terms of the gradual accumulation of small changes as postulated by natural selection. In fact, historians often refer to the "eclipse of Darwinism" in the early decades of the twentieth century. Then, in the 1930s and 1940s, brilliant work by mathematicians and scientists such as R. A. Fisher, Sewell Wright, Theodosius Dobzhansky, Ernst Mayr, and G. G. Simpson established the neo-Darwinian synthesis. The synthetic theory showed that, instead of being opposed, Darwinian and Mendelian theory needed and mutually supported each other. In fact, researchers found that the way to understand the genetics of populations was to incorporate models of Darwinian selection. This synthesis was a triumph for both theories. United into an integrated synthetic theory, both Darwinian evolutionary theory and Mendelian genetics were made stronger than ever. In fact, evolutionary theory connects at such a fundamental level with so much in biology that Dobzhansky famously said, "nothing in biology makes sense except in the light of evolution."

The history of science contains many incidents where seemingly disparate theories, even theories in very different fields of science, are found to be mutually supportive. Perhaps, then, we can view such episodes as supporting, at the very least, a minimal form of realism, what philosopher Robert Almeder calls "blind realism." That is, even if we are not confident that any *particular* theory is an approximation of reality, it seems simply inconceivable that the *whole* edifice of scientific theory could be just wrong. Likewise, if we have already filled in a good bit of a crossword puzzle, it could well be that some of our answers are wrong, but it seems highly unlikely that *every* answer is wrong! Therefore, given the history of greatly increasing empirical success and the vast increase in the cohesiveness and mutual supportiveness of our theories, it just does not seem feasible that we could be completely on the wrong track. If anyone insists that, nevertheless, we might be completely wrong, we can rightly demand that they give us a reasonably detailed picture of what a total alternative science looks like.

At this point many antirealists would complain that the above argument is directed at a straw man. (I do not think so. I think that global antirealism is quite common, even faddish, these days.) They would object that they do not assert that the *whole* body of scientific theory could be wrong, merely that particular kinds of theoretical truth-claims cannot be known to be true. Perhaps they would be perfectly willing to admit that theories in evolutionary biology, paleontology, geology, and other fields that postulate measurable, detectable events and processes are perfectly fine with them. Their problem is with theories, like those of fundamental physics, that postulate entities that are *in principle* impossible to directly see or measure. Things and occurrences that are merely unobservable *in practice*, like an asteroid strike 65 million years ago, give them no problems (such an event would have been eminently observable to any sentient creature unlucky enough to have been around when it occurred). Their objection is that science so often postulates entities, like electrons, that cannot possibly be directly observed. In cases like this, they say, we go farther than we should in asserting that such things really exist. In the next chapter we examine the claims of this more modest antirealism.

FURTHER READINGS FOR CHAPTER FOUR

Herbert Butterfield coined the term "Whig history" in *The Whig Interpretation of History* (New York: Charles Scribner's Sons, 1951). In another work he gives a very succinct statement of the problems with such historical writing:

> It is not sufficient to read Galileo with the eyes of the twentieth century or to interpret him in modern terms—we can only understand his work if we know something of the system which he was attacking, and we must know something of that system apart from the things which were said about it by its enemies. In any case, it is necessary not merely to describe and expound discoveries, but to probe more deeply into historical processes and learn something of the interconnectedness of events, as well as to exert all our endeavours for the understanding of men who were not like-minded with ourselves. Little progress can be made if we think of the older studies as merely a case of bad science or if we imagine that only the achievements of the scientist in very recent times are worthy of serious attention at the present day.

This quote is from pages 9–10 of Butterfield's *The Origins of Modern Science 1300–1800, revised edition* (New York: Free Press, 1957). A good discussion of the problems with Whig history is in Chapter Nine of *An Introduction to the Historiography of Science* by Helge Kragh (Cambridge: Cambridge University Press, 1987).

As I note in the Further Readings for Chapter Two, the work that set the record straight on Owen was Adrian Desmond's *Archetypes and Ancestors: Palaeontology in Victorian London 1850–1875* (Chicago: University of Chicago Press, 1982). Desmond shows that Huxley was the vanguard of a movement to completely secularize science and rule out the appeal to metaphysical principles like archetypes. Owen, on the other hand, represented the more conservative elements of Victorian society who were shocked by the materialism and agnosticism of the young scientific radicals.

Steven Shapin and Simon Schaffer's *Leviathan and the Air-Pump* (Princeton: Princeton University Press, 1985) retrieves a rather obscure controversy from the history of science and turns a seemingly dry subject into a fascinating narrative. The authors certainly succeed in showing, as did Desmond in the work mentioned above, that seemingly arcane scientific debates are about a lot more than they seem. The fact that scientific issues (like global warming in our day) can become footballs for conflicting political doctrines does create a severe challenge for the autonomy and objectivity of science. For a dissenting view on the Hobbes/Boyle controversy, see the third chapter of Paul R. Gross and Norman Levitt's *Higher Superstition: The Academic Left and Its Quarrels with Science* (Baltimore: Johns Hopkins University Press, 1994). See also the eleventh chapter of Christopher Norris's *Against Relativism: Philosophy of Science, Deconstruction, and Critical Theory* (Oxford: Blackwell Publishers, 1997).

Paul Feyerabend long relished his role as "the worst enemy of science." His book *Against Method* (London: NLB, 1975), was a strident attack on "scientific method." Feyerabend is best characterized as an extreme libertarian; he insisted that no perspective or ideology should enjoy any sort of hegemony. Science should therefore be demoted from its dominant position in Western society and should have no higher status (or more funding) than any other way of looking at the world. Astronomy, for instance, should enjoy no higher prestige or support than astrology. The idea that science is a more "objective" or "rational" way of interpreting reality than, say, voodoo, was loudly and contemptuously derided by Feyerabend. A sympathetic but critical study of Feyerabend is John Preston's *Feyerabend* (Cambridge: Polity Press, 1997).

Larry Laudan's insightful remarks about scientific progress are found in his essay "The Philosophy of Science and the History of Science" in *A Companion to the History of Modern Science,* edited by R. C. Olby, G. N. Cantor, J. R. R. Christie, and M. J. S. Hodge (London: Routledge, 1990), pp. 47–59. Laudan's case against convergent realism is given in his essay "A Confutation of Convergent Realism" in *Scientific Realism,* edited by Jarrett Leplin (Berkeley: University of California Press, 1984), pp. 218–249, and in his book *Science and Values: The Aims of Science and Their Role in Scientific Debate* (Berkeley: University of California Press, 1984). Laudan's arguments against scientific realism are clear and challenging. It is an especially effective critique because it is grounded in deep knowledge of the history of science. A powerful riposte to Laudan is given by Stathis Psillos in his book *Scientific Realism: How Science Tracks Truth* (London: Routledge, 1999). Psillos's book is perhaps the most comprehensive and effective defense of scientific realism currently available. Jarret Leplin's *A Novel Defense of Scientific Realism* (Oxford: Oxford University Press, 1997), is also a vigorous and carefully argued statement of scientific realism and critique of antirealist arguments.

Among the writers of science for the general public, some, like Carl Sagan, could compose beautiful, resonant prose that conveyed the wonder and majesty of the cosmos. Others, like Stephen Jay Gould, produced writings that were masterpieces of the essayist's art. But nobody could match Isaac Asimov in sheer expository skill. Asimov had the very rare, perhaps unique, ability to take the most difficult ideas of science and present them so clearly and engagingly that any interested person could read about them with pleasure and profit. Asimov seldom made comments of philosophical significance, but his delightful essay "The Relativity of

Wrong" is an exception. It is printed in one of many collections of Asimov's essays, *The Relativity of Wrong* (New York: Kensington Books, 1988). Asimov makes the simple point—one that seems to have eluded many philosophers—that while there may be only one absolutely correct way to answer a question, there are infinitely many ways to answer it falsely, and some false answers reveal far more insight about the topic of the question than others. Maybe philosophers would do better to spend less time trying to develop theories of approximate truth and more time exploring the cognitive value of insightful falsehoods.

An authoritative and highly readable history of atomism is Bernard Pullman's *The Atom in the History of Human Thought*, translated by Axel Reisinger (Oxford: Oxford University Press, 1997). Pullman traces the ups and downs of atomic theory through the centuries and shows how it finally triumphed only at the beginning of the twentieth century.

W. V. O. Quine expresses his insights into the interconnectedness of all our assertions, from the most theoretical to the most mundane, in many places in his writings. A very clear and succinct statement is in the "Introduction" to his *Methods of Logic*, fourth edition (Cambridge: Harvard University Press, 1982). For an elementary exposition of his view of rational belief, see *The Web of Belief*, second edition, coauthored with J. S. Ullian (New York: Random House, 1978). The "web of belief" is Quine's main metaphor for the complex connections between our beliefs and the way, like an extremely intricate spider web, that different strands support, and are supported by, many other strands. Susan Haack makes a similar point with her crossword puzzle analogy. She develops the analogy and offers many other insights in a book I recommend to the reader with no reservations, *Defending Science—Within Reason: Between Scientism and Cynicism* (Amherst, N.Y.: Prometheus Books, 2003). In several of her works Haack defines herself as a "passionate moderate" who rejects both the immoderate claims of scientism and the cynicism of the radical science critics.

A work I failed to mention earlier that gives an excellent overview of the history of evolution, including the evolutionary synthesis, is Peter J. Bowler's *Evolution: The History of an Idea*, third edition (Berkeley: University of California Press, 2003). Bowler is one of the most prolific and knowledgeable historians of Darwinism and evolutionary ideas. An excellent history of Gregor Mendel's discoveries is Robert Olby's *Origins of Mendelism* (Chicago: University of Chicago Press, 1985).

Robert Almeder's theory of "blind realism" is presented in his *Blind Realism: An Essay on Human Knowledge and Natural Science* (Lanham, Md.: Rowman and Littlefield, 1991). Almeder's arguments are sharp and crystal clear. His views on realism have not received the attention they deserve.

5

TRUTH OR CONSEQUENCES?

SEEING IS BELIEVING, WE ARE TOLD. About the best evidence we can
have for anything is to see it. What would clinch the case for UFOs? If
one landed on the Washington Mall, as in the science fiction classic *The
Day the Earth Stood Still,* that would clinch it. What would prove that the
Loch Ness Monster is real? If the body washed up tomorrow. What would
fill churches, synagogues, and mosques with former unbelievers? Atheist
philosopher Norwood Russell Hanson imagines the scenario that would
do it for him:

> Suppose . . . that on next Tuesday morning just after our breakfast all of
> us in this one world are knocked to our knees by a percussive and ear-
> shattering thunderclap. Snow swirls; leaves drop from the trees; the earth
> heaves and buckles; buildings topple and towers tumble; the sky is ablaze
> with an eerie, silvery light. Just then, as all the peoples of this world look
> up, the heavens open—the clouds pull apart—revealing an unbelievably
> immense and radiant Zeus-like figure, towering up above us like a hun-
> dred Everests. He frowns darkly as lightning plays across his Michaelan-
> geloid face. He then points down—at me!—and exclaims for every man,
> woman, and child to hear: "I have had quite enough of your too-clever
> logic-chopping and word-watching in matters of theology. Be assured,
> N. R. Hanson, that I do most certainly exist." (Hanson, 1971, p. 309)

Yes, it would be hard to remain a skeptic in circumstances like these,
but what if the evidence for God's existence is not this good? When is
it reasonable to believe in God even without such spectacular Steven
Spielberg-type revelations? In general, when should we believe in something we cannot see? Well, maybe if we could hear, taste, smell, or feel it.

143

But what if something cannot be detected by any of the senses? As a matter of fact, people believe in all sorts of things that cannot be detected by any of the senses. Many people believe that gods, souls, evil spirits, luck, fate, numbers, *mana, ch'i,* and so on, exist even though they are not objects that can register on the senses. It is very important to ask how we can have confidence that things exist even if we cannot sense them.

A very common misconception about science is that it deals only with the observable. A common anti-evolutionary charge is that since we cannot see creatures evolve from one major group to another (reptiles to birds, say), it is not scientific to claim that they did. On the contrary, scientists constantly talk about things that are too big, too small, too fast, too slow, too inaccessible, or too distant in space or time to see. If we had time machines we could do paleontology by traveling back to the past to directly observe prehistoric life. Of course, there would be problems with that too, as Gary Larson showed in his *Far Side* cartoon of a time-traveling scientist approaching the backside of a huge dinosaur with an oversized rectal thermometer. The caption read: "An instant later, both Professor Waxman and his time machine are obliterated, leaving the coldblooded/warmblooded dinosaur debate still unresolved." But since we have no time machines, we cannot observe past life directly and have to draw our inferences from the traces we have now. Also, many fields of science talk about things that cannot be directly sensed—extrasolar planets, magnetic fields, electrons, and so forth—but are known only through their effects on various sorts of detection devices (spectrometers, electron microscopes, Geiger counters, seismographs, etc.).

Here we need to make an important distinction. There are some things we cannot see, but this is only a *practical* problem. We cannot travel to the earth's inner core, so our knowledge of the core is inferential, not based on direct observation. However, if we *could* devise a way to burrow through thousands of miles of crust, mantle, and outer core we could directly sample the earth's inner core. The problem is that no conceivable technology would ever permit us to do that. So, the earth's inner core is unobservable—but only *in practice,* not *in principle.* Contrast this with the case of the electron. Even if (to cite another 1950s science fiction classic) like *The Incredible Shrinking Man* you could be reduced to atom size, you still would not be able to see electrons. An electron is far smaller than the wavelength of light, so electrons are *in principle* invisible (the in practice/in principle distinction is not really very clear, but let's go with it for now). Yet, for over a hundred years, physicists have accepted the existence of electrons. To see why they did, and still do, regard electrons as genuine entities, we need to review the history of their discovery.

ELECTRONS: REAL PARTICLES OR CONVENIENT FICTIONS?

Electrical phenomena have long fascinated people. Since ancient times it has been known that rubbing certain substances like amber (the word for "amber" in Greek is "*electron*") would give them the strange power to attract small bits of paper and other materials. It was also known that this strange attractive power could be passed by direct contact from one body to another. In the eighteenth century scientists demonstrated that the electrical effect could be transmitted long distances if you used the proper conductive medium. So, it looked like *something* is transferred from electrically charged bodies to uncharged ones. Scientists postulated a subtle electrical fluid (others favored two fluids) that could be agitated by friction and could flow along wires and from one body to another. However, two of the leading researchers of electricity in the nineteenth century, Michael Faraday and James Clerk Maxwell, disagreed with the subtle fluid theory and maintained that electrical charge is a "field" phenomenon, a tension in the medium surrounding electrified bodies.

Another hypothesis favored by some early researchers was that electricity was the effect of the interaction of electrical corpuscles—tiny, irreducible particles bearing the smallest units of electrical charge. Some physicists regarded such electrical particles as fundamental constituents of matter. With the rise of Maxwell's field theory, these hypotheses were largely abandoned, but they soon came back in a different guise. In 1891 Irish physicist George Johnstone Stoney introduced the term "electron" as a measure of unit electrical charge. Also in the early 1890s the noted Dutch physicist Hendrick A. Lorentz attempted to reconcile the electrical particle hypothesis with the Maxwellian field theory by proposing that electrons were structures in the continuous medium postulated by Maxwell. Clearly, this was a time of much ferment and little consensus among physicists in their thinking about electricity. In his classic of irreverence *The Devil's Dictionary*, Ambrose Bierce, the leading American satirist of the late nineteenth century, mocked physicists' apparent indecisiveness about electricity. His tongue-in-cheek definition of "electricity" was "The power that causes natural phenomena not known to be caused by something else." In other words, in Bierce's view, physicists were clueless about the nature of electricity. However, just when it looks to outsiders like scientists are grasping at theoretical straws, in fact they are often on the cusp of a major advance.

Progress often occurs in science when there is a fruitful interaction of theory and experiment. While the theorists were busy proposing and debating hypotheses, a particular problem had grasped the attention of experimentalists: What are cathode rays? If you take a glass tube and pump enough air out to make a reasonably good vacuum, attach a negative electrode (the cathode) and a positive electrode (the anode) to opposite ends of the tube, and send a strong electrical current into the tube, an electrical discharge will flow from the cathode through the tube to the anode. So, *something* flows from the cathode to the anode, but what is it? Is it a "ray"—that is, an electromagnetic wave—or a stream of tiny charged particles?

In 1897 English physicist J. J. Thomson, head of the famous Cavendish Laboratory at Cambridge University, performed an experiment that strongly supported the theory that cathode "rays" are particles. Thomson found that an electrical field could deflect the "rays," something that would be expected of charged particles, but not electromagnetic waves. Further, he was able to determine the ratio of mass to charge that the hypothesized particles would have to possess. In an earlier experiment he had shown that the velocity of the "rays" was far less than the speed of light, which also undermined the electromagnetic wave theory. Putting all the pieces of evidence together, Thomson declared that the cathode "rays" are in fact minute particles bearing a negative electrical charge. The name "electron" was eventually given to the postulated particle, and Thomson is usually given the main credit for its discovery.

It is remarkable that at about the same time Thomson was doing his experiment, German physicist Walter Kaufmann was doing the same kind of experiment in Berlin, only, according to some historians, he did it *better.* Yet Kaufmann is not credited with the discovery of the electron because he did not claim to have discovered a new particle. Kaufmann was not less familiar with the physics than Thomson; his reluctance to proclaim the existence of a new particle was due to a *philosophical* conviction. The philosophy called "positivism" was very influential among German scientists at the time, largely due to the influence of scientist/philosopher Ernst Mach (1838–1916). The basic principle of positivism is that scientists have the job of systematically describing only what can be measured and should not speculate on the existence of things that cannot be directly detected. For Mach, a scientific theory does not function to reveal to us the hidden nature of reality; it is an *instrument* to enhance our powers of prediction and control over what we can see. Mach therefore had what is called an "instrumentalist" view of theories. Of course, Mach realized that scientists sometimes find it useful to speak of invisible things like atoms or fields, but he held that such terms are used only as figures of speech and are not meant literally. Concepts like "atom" and "field" are nothing more than convenient fictions, says Mach; they are placeholders for batches of observations or measurements that we could perform but are too numerous to mention individually. For instance, when we say that a magnetic field exists in a certain area of space, this is just a shorthand way of saying that, for instance, iron filings scattered in that space would arrange themselves in a certain way, or that a compass needle would be deflected in a certain direction, and so forth.

The appeal of positivism is obvious. To many working scientists of the day it seemed that the chief advantage science possesses over metaphysics is that science sticks to observable facts and refuses to engage in airy speculation about inscrutable entities. Yet positivism arguably impeded the progress of science, and was abandoned, even by some of its most ardent proponents, when the evidence for the existence of unobservable entities became overwhelming. In his 1913 book *Les atomes*, physicist Jean Perrin gave the strongest evidence yet obtained for the reality of the world of atoms and molecules. Every beginning chemistry student is introduced to Avogadro's number (named after Italian chemist Amedeo Avogadro). This is the number of atoms in the gram molecular weight of any substance. To get the molecular weight of any substance, first add up the atomic weights of the atoms that make up the substance (for instance, a molecule of water has a molecular weight of about 18 since each molecule of water is two atoms of hydrogen and one atom of oxygen, and each atom of hydrogen has an atomic weight of about 1 while the oxygen atom has a weight of around 16). Now take the number of grams of the substance equal to its molecular weight—eighteen grams in the case of water. Eighteen grams of water will contain Avogadro's number of water molecules. Avogadro's number is very large—6.022045×10^{23}—and this gives you some idea of how tiny water molecules are since 18 grams is not much water. Perrin cited thirteen independent ways of determining Avogadro's number, and they all gave very similar values. When thirteen independent methods arrive at pretty much the same answer, it is very hard for even the most skeptical scientist to doubt that there is something real underlying these results. In other words, when we use thirteen completely different procedures to count the number of atoms in a given amount of stuff, and we get close to the same number each time, it is hard to believe that we are not counting *something*.

It surely is much harder to doubt the existence of molecular reality today than it was when Perrin wrote *Les atomes* in 1913. After all, we now have the rapidly growing field of nanotechnology, where we actually build and use devices on a molecular scale. We now make micromachines, like gears that are smaller in diameter than a human hair. Scanning electron microscopes can apparently see and manipulate individual atoms and molecules (more on this later in the chapter). Such devices can also be used for ultrafine etching. You could engrave the entire *Encyclopedia Britannica* on the head of a pin. Scientists and engineers engaged in such work would likely feel little patience for a skeptic who still doubted the existence of atoms and molecules. Their response to such skepticism would likely be to point to the images on the screens of their electron microscopes and exclaim with exasperation, "But, dammit, there they are!" But skeptics do still exist, and since they are among the most prominent figures in contemporary philosophy of science, we need to look at their arguments.

VAN FRAASSEN'S CONSTRUCTIVE EMPIRICISM

In 1980 noted philosopher of science Bas van Fraassen published his book *The Scientific Image.* Although this book is not as famous among non-philosophers as Kuhn's *The Structure of Scientific Revolutions,* it has had a strong impact on the philosophy of science. Van Fraassen defends a form of antirealism he calls *constructive empiricism* (CE). CE is a more moderate form of antirealism than the instrumentalist view of Mach. As we saw, Mach held that scientific theories are not even *about* the deep, hidden structure of the universe; all they are about is the prediction and control of observable things. Van Fraassen, on the other hand, takes theories literally. In his view, theories about atoms *really do* postulate the existence of such invisible entities and do not merely use the term "atom" as a figure of speech. The same holds for all theories that propose the existence of "theoretical entities," things that are unobservable but which are supposed to explain the phenomena we do observe. Therefore, for van Fraassen, theories that postulate theoretical entities are intended to be literal descriptions of the hidden structure of the universe and what they assert is either true or false. Van Fraassen opposes "scientific realism," which he understands as follows: "[According to scientific realism,] science aims to give us, in its theories, a literally true story of what the world is like; and acceptance of a scientific theory involves the belief that it is true." (Van Fraassen, 1980, p. 8)

A number of scientific realists have objected that van Fraassen has given a misleading and simplistic characterization of their position. In particular they object that he has construed their view as a thesis about the *truth* of theories when scientific realism is really an assertion about the *objective existence* of the kinds of entities postulated by our physical theories (Devitt, 1991). However, for the sake of convenience, and to avoid muddying the waters with too many caveats, qualifications, and distinctions, let's go with van Fraassen's definition for now. A more adequate characterization of scientific realism will be developed at the close of this chapter.

Van Fraassen proposes an alternative to the realist view of the aim of science. He holds that science would be just as worthwhile and valuable an activity if we took the more modest view that the aim of theory is merely to achieve empirical adequacy. An empirically adequate theory is one that makes predictions consistent with all past, present, and future observations. Van Fraassen certainly thinks that we should *accept* the best-supported current theories, but he holds that accepting a theory is not the same thing as believing it. For instance, you can accept the reigning standard model of particle physics without believing that the particles it postulates (quarks, electrons, etc.) are real. Van Fraassen recommends that, with respect to all well-confirmed theories that postulate unobservable entities, we should accept those theories without believing them. He says that by accepting (but not believing) such a theory you commit yourself *only* to the belief that its observable predictions have been and will continue to be reliable. In other words, Van Fraassen is reviving the view that the goal of science is not to discover truth but merely to save the appearances.

So, what do we want from scientific theories—that they be true depictions of reality or merely that their observational consequences be accurate? Let's begin by looking at van Fraassen's argument in some detail. First we need to get clear on just how strong his claim is. Is he saying that with respect to theories that postulate unobservable theoretical entities, we can never rationally believe that the theory is true? That is, is he asserting that no matter how well such a theory is confirmed, or how completely it meets our theory-choice standards, it is never reasonable to believe that the theory is true? His claim is milder than this, at least as he expresses it in some of his writings. He merely claims that realism is not rationally *compelling*, that is, that an antirealist attitude towards theories, and concomitant agnosticism about the existence of theoretical entities, is just as reasonable as the realist attitude. In other words, antirealism is no less rational a view of scientific theories than realism.

Actually, put this way, van Fraassen's antirealism is so mild a doctrine that it hardly seems to merit the vast amount of commentary and argument it has evoked. Realism need not be seen as a form of intellectual imperialism, though some of the radical science critics mentioned in Chapter Three might think that it is. Even the most hard-bitten realist need not insist that it is downright *irrational* not to believe in electrons, quarks, or whatever. After all, some of the leading physicists have entertained doubts about theoretical entities. Most famously, Niels Bohr, one of the leading physicists of the twentieth century and one of the founding figures of quantum mechanics, took the frankly antirealist view that the goal of physics was the correct prediction of experimental results, not to penetrate to an invisible reality behind the appearances. He said, "In our description of nature, the purpose is not to disclose the real essence of phenomena, but only to track down, as far as possible, the relations between the manifold aspects of our experience" (quoted in Kragh, 1999, p. 210). Werner Heisenberg, another founder of quantum mechanics, agreed with Bohr. It would be bold indeed to say that physicists of the caliber of Bohr and Heisenberg held an irrational view about the nature and goal of physics. So, if all that van Fraassen is claiming is that his antirealism is not an irrational position—while also admitting that realism is a fully reasonable view—there seems to be no reason to belabor the issue and the point can just be conceded.

Yet there are places where van Fraassen gives arguments that apparently support a somewhat stronger claim. That is, some of his arguments seem to uphold the view that it is *less* reasonable to believe a well-confirmed theory (i.e., regard it as true) than merely to accept it as empirically adequate. For instance, while he admits that the unobservable entities scientists postulate *may* exist, van Fraassen contends that we can never *know* that they do. In other words, he is not claiming that there is no reality below the level that we can observe. Rather, he agrees with scientific realists that there is such a reality, but he contends that we cannot be sure what that reality is like (in philosophical jargon, van Fraassen's claim is epistemological, not metaphysical). No matter how much evidence we might have supporting the existence of, say, electrons, it is always possible, since we cannot see the electrons, that there is something else there that is not an electron but just mimics the way we think electrons ought to act.

Also, van Fraassen seems to make a rather strong claim about the kind of observational knowledge science can give us. He puts a great deal of stock in what we can observe with the unaided senses. He does not deny that our observations are "theory laden," that is, that the content of our observational knowledge is pervasively and fundamentally dependent on the general theories we accept. Even the most mundane observation, like "this is a glass of water," is not just a report of raw experience, but involves the ideas "glass" and "water" which are general concepts that we use to *interpret* our experience. Nor does van Fraassen think that there is a difference in kind between statements that assert observational claims and those that assert theoretical claims (earlier philosophers of science made a strict distinction between "observation language" and "theory language"). However, he does think that, as a matter of fact, some objects are observable, while some hypothesized objects are in principle unobservable. That is, some of the objects postulated by theories are such that organisms of our biological sort could not observe them no matter where we were located or when we might exist. Van Fraassen counts as "observable" things like Charon, the moon of Pluto, that are just too far away for us to see, or events, like the impact that supposedly eradicated the dinosaurs, that happened long before there were any humans. Such objects and events still count as observable for van Fraassen because could we be transported to Pluto, or had we existed at the end of the Cretaceous, we could have observed those things. Only things that are observable, in this broad sense, are objects of empirical knowledge for van Fraassen. He is skeptical about the claims scientists make about things that are in principle unobservable, even if instruments like microscopes supposedly detect them. It is not that he denies the existence of things too small to see, only that he thinks that we cannot know the microworld as we know the world we can observe with the unaided senses.

I therefore take van Fraassen to be making two claims about scientific knowledge: (1) Our observational knowledge extends only so far as our unaided senses can, in principle, go. A considerable degree of skepticism is warranted with respect to what supposedly can only be detected by instruments. (2) When we accept a scientific theory that postulates unobservable entities, we should commit ourselves *only* to its empirical adequacy and not make the stronger and riskier claim that it is *true*. As we noted in the last chapter, false theories can make true predictions, so a false theory might prove as empirically adequate, even in the long run, as a true theory. When we accept a theory, therefore, it looks like a safer bet if we commit ourselves only to its empirical adequacy and suspend belief about its truth. Why stick your neck out any further than you have to? In the next section I shall examine van Fraassen's claims about observational knowledge, and in the following section move on to examine his skepticism about theories that postulate unobservable entities.

DO WE OBSERVE THROUGH MICROSCOPES?

How much trust should we place in our unaided senses? Is there any reason to think that the operations of our sensory faculties are, in principle, more reliable than other ways of getting information about the world? We have to be very careful here. Just because we are more familiar with using our eyes than, for instance, an electron microscope does not mean that our eyes are less likely to convey false or misleading information. The fact that we use our eyes naturally and generally without any special training, makes us feel that seeing is a simple and unproblematic process.

But seeing is neither simple nor unproblematic. For one thing, the human visual apparatus, consisting of eyes, the optic nerve, and the portions of the brain that process visual input, constitutes a staggeringly complex system (or, rather, a system of systems of systems of . . .). Compared to the human visual apparatus, a machine like an electron microscope is really a pretty simple device. In fact, it is fair to say that neuroscientists have only in very recent years begun to put together anything like an adequate account of how the brain processes visual input (this story is engagingly told in the late, great Francis Crick's *The Astonishing Hypothesis*). Much, of course, is still unknown. By contrast, the physical principles on which the electron microscope operates are as well known as any in physics.

More to the point, a vast scientific literature exists detailing how sense perception is *not* reliable in many circumstances. Consider UFO sightings. Psychologist Terence Hines describes some of the visual illusions that have contributed to many UFO "sightings":

> A number of well-known visual illusions play a role in what witnesses report in UFO sightings, especially those that take place at night. One is the *autokinetic effect.* This effect refers to the fact that, if one views a small source of light in a dark room, the light will appear to move, even though it is stationary, and even though the observer's head is stationary. . . . Another illusion is that of *apparent motion.* Consider two positions in a dark room, A and B. A small light is turned on at A, then turned off. Moments later, a second light is turned on at B. . . . What is perceived is a single light appearing at point A, moving to point B, and then going off. (Hines, 1988, pp. 170–171)

Even more interesting are the ways that people can "see" what they expect to see, or what they want to see, or what they think they ought to see rather than what actually occurred. Numerous psychological experiments have demonstrated how personal expectations and social pressures bias perception; no doubt this is the source of many reports of "miraculous" and "paranormal" happenings. Some local news programs tout the reliability of their reporting by naming their broadcasts "Eyewitness News." Any competent trial lawyer can quickly show how fallible eyewitness testimony can be.

The point is not that we should never trust our senses (after all, what choice do we have?). If we could never trust our senses, we would have no basis for believing those sciences that tell us that our senses are not always to be trusted! The point is that observation is highly fallible, easily misled, and often subject to revision or correction. Indeed, observation is often far more reliable when we supplement our unaided senses with the use of instruments. *Prima facie,* then, there is no reason to think that the unaided senses are in general more trustworthy than machines, like an electron microscope, specifically designed to show things beyond the reach of human senses.

Why do scientists trust what they see through microscopes? How do we know that what microscopes show us is real and not just an artifact, a false image produced by the microscope itself rather than the actual features of the specimen we are supposedly observing? Well, for one thing, many of the things we can see only through microscopes look really real! If you collect a tiny bit of detritus from your bedroom carpet and put it under a good microscope at about 1000 power, you will see—or, in deference to van Fraassen, you will *seem* to see—fearsome-looking creatures. Biologists, who believe that these creatures are real and not figments of microscopy, tell us that these creatures are dust mites, tiny arachnids (related to spiders, that is) that live in your carpets and eat the skin flakes you constantly shed. The microscope will apparently reveal a great deal of detail about this creature. It seems to have eight legs, serrated front claws, protective plates of body armor, and holes for breathing, eating, excretion, and copulation. Again, it looks very real. I would expect an artifact, a false image produced accidentally by the operation of the instrument, to look like a blur or a blob, not something that the computer animators at Pixar Studios would have been proud to create. In fact, microscopes seem to reveal an entire microworld, one as rich in detail and as diverse in population as our everyday world. To take seriously the suggestion that all of this could be an artifact, we would have to be given a very convincing explanation of just how microscope makers could have created this world within the world.

Another reason to trust some of the images microscopes give is that some imaging processes can display a continuous range of magnifications, from just below what the unaided eye can see to very high powers. For instance, there is a well-known series of micrographs that begins with an image of the tip of a pin that is magnified slightly. You can easily see the smoothly tapering tip. Also, there are tiny yellow clusters on the tip, but you can't quite make out what they are. In the next image you can still see the very tip of the pin and the yellow clusters have begun to resolve into piles of tiny individual shapes. Closer views show that these piles are composed of fuzzy, yellow, lozenge-shaped bacteria that find a home in the niches of the pin. What you have then is a series of pictures that connect at one end with what the unaided eye can see. Successive magnifications resolve and bring into focus features that were visible but less clearly seen in the previous images. In short, you have a series of pictures, continuous with ordinary vision, that display just what you would expect to see if you were getting closer and closer views of the subject.

Whether we actually see through a microscope is a question philosopher Ian Hacking raises in one of the most acute criticisms of van Fraassen's skepticism about the reliability of instruments designed to see what the human senses cannot detect. Hacking's essay "Do We See Through a Microscope?" begins by reviewing the history of microscopy from Leeuwenhoek's first primitive device in the seventeenth century, little more than a toy, to today's sophisticated instruments. Hacking shows how refinements in lenses removed the distortions that made early microscopes practically worthless for serious research. In the late nineteenth century Ernst Abbe, part-owner of the now world-famous Zeiss optical firm, succeeded in mating theory and practice to produce light microscopes of superior resolution. In the twentieth century, microscopes based on various different physical properties of light, as well as different kinds of electron microscopes, were developed. As microscopes advanced, they became of greater and greater use and finally indispensable to the practice of many branches of science.

Hacking says that one reason to think that what microscopes show us is real is that the physical principles, the laws of optics, used in the construction of microscopes are very well understood. However, Hacking says that this is only a minor reason why we can trust the images of microscopes. A better reason is that we can physically interact with the structures that we see through the microscope. For instance, we can watch as we insert tiny needles into cells and see the liquid ooze out as we inject it. Thus, we do not just passively look through a microscope; we can use it to do work in the microworld, moving around and intervening in that world. We noted earlier that there are now electron microscopes so powerful that they apparently allow us to manipulate individual molecules. Surely, if we can push things around, build things with them, or inject stuff into them, they must be real.

Perhaps the most persuasive evidence of the reality of the denizens of the microworld is that different microscopes, designed according to entirely different physical principles, show us pretty much the same thing. In fact, we can manufacture grids, label each grid with a letter, and shrink the grids to microscopic size. When we look through a microscope, any kind of microscope, we can still see the grid and the little letters in them. Hacking therefore challenges the antirealist:

> Can we entertain the possibility that, all the same, this is some gigantic coincidence? Is it false that the disc is, in fine, in the shape of a labeled grid? Is it a gigantic conspiracy of 13 totally unrelated physical processes that the large scale grid was shrunk to some non-grid which when viewed using

12 different types of microscopes still looks like a grid? To be anti-realist about the grid you would have to invoke a malign Cartesian demon of the microscope. (Hacking, 1985, p. 147)

René Descartes imagined that there could be an all-powerful demon that so deceived us that we would go wrong no matter how obviously something seemed to be so. Hacking says that in order to remain skeptical about the reality of the things we seem to see through microscopes we would have to postulate such a demon.

Hacking's essay was written more than twenty years ago. Subsequent developments in nanoscience have greatly strengthened his case. Nanoscience studies phenomena at the "nano" scale; a nanometer is one-billionth of a meter, far smaller than anything the unaided eye can see. Various devices have been developed to study reality at these scales. Two leading nanoscientists, Mark Ratner and Daniel Ratner, describe the use of "scanning probe" instruments, comparing their operation to the way that we find out about surface textures by sliding a finger over them:

> In scanning probe instruments, the probe, also called a tip, slides along a surface the same way your finger does. The probe is of nanoscale dimensions, often only a single atom in size where it scans the target. As the probe slides it can measure several different properties, each of which corresponds to a different scanning probe measurement. For example, in *atomic force microscopy* (AFM), electronics are used to measure the force exerted on the probe tip as it moves along the surface. This is exactly the measurement made by your sliding finger, reduced to the nanoscale. (Ratner and Ratner, 2003, pp. 39–40)

In principle, "feeling" at the nanoscale is analogous to feeling with the finger. In practice, of course, computers must be used to enhance the signal from the tiny probe, and this is a fairly complex process. But again, feeling with the human finger is a far more complex process and is not as well understood as the physical principles on which atomic force microscopy operates. From the perspective of our best-confirmed scientific theories, which is the perspective van Fraassen endorses, there seems to be no reason to think that atomic force microscopy is any less reliable in conveying information about the things it "feels" than our fingers are when we detect the roughness of sandpaper or the smoothness of velvet.

Further, nanoscience has evolved into nanotechnology, that is, we now have the tools to build structures at the nano scale that have the properties we design into them, properties with effects detectable at the "macro" scale. The most obvious examples are computer microchips. As of February 2002, when the Ratners were writing, the smallest features commonly etched onto commercial microchips were about 130 nanometers across. The diameter of a human hair is 50,000 nanometers, and the smallest things the unaided eye can see are 10,000 nanometers wide. Given our ability to build things at the nanoscale that reliably operate the way they are designed to operate, it looks like we have a pretty good idea of what is going on down there. If we don't know what we're doing in making a microchip, then our efforts to design and manufacture them are just a shot in the dark, and it is amazing not just that we can make them, but that we are rapidly getting better and better at it!

Let me pause to emphasize that van Fraassen does not maintain the patently absurd view that there is no microworld or that all of our images of that world must be artifacts or fabrications. However, he does hold that our beliefs about things that, if they exist, are detectable only through instruments cannot be nearly as solidly grounded as our beliefs about things we can directly sense. He also opposes arguments like Hacking's that seem to be intended to show that skepticism about microscopic reality is simply perverse. In his reply to Hacking, van Fraassen focuses on the charge that so many different kinds of microscopes, operating on very different physical principles, would not show the same thing unless what they showed was real:

> Imagine I have several processes that produce very different visual images when set in motion under similar circumstances. I study them, note certain similarities; as I repeat this, I discard similarities that do not persist and also build machines to process the visual output in a way that emphasizes and brings out the noticed persistent similarities. Eventually the refined products of these processes are strikingly similar when initiated under similar circumstances. Now I point to the similarities and say that they are too striking to be there by coincidence, though, of course, the discarded dissimilarities were mere idiosyncrasies of the individual processes. What is the status of my assertion? What principle of reasoning could support it? Since I have carefully selected against nonpersistent similarities in what I allow to survive in the visual output processing, it is not at all surprising that I have persistent similarities to display to you. (van Fraassen, 1985, pp. 297–298)

In other words, it could be the microscope makers, not the supposedly objective features of the examined specimen, that account for the agreement in what is seen. That is, a microscope is deemed a reliable instrument only when it has been adjusted to show images like those of other microscopes. But perhaps building microscopes is just a case of "the blind leading the blind," to use the Biblical metaphor, and microscopes have been unwittingly designed to display the same false images.

However, this is a weak argument that does very little to undercut Hacking's point. Hacking's point is that it is very unlikely that different microscopes, designed to operate on very different physical principles, would show closely similar images unless those images were real. Van Fraassen's reply would be cogent only if microscope designers, wittingly or not, *constructed* the image to look like the images from another microscope. Merely *selecting* among various images will not do; you can only select from what is there. Neither can you refine or emphasize the *similarities* between images unless the similarities are there. Van Fraassen offers no argument for thinking that microscope makers construct factitious images rather than select and refine striking similarities that are *already there.*

It looks like a good cumulative case can be made that we do *observe* things with microscopes that are really there but too small to be seen with the unaided eye: Many of the things we seem to see look very real, some with legs and other organs very much like the ones we see on large-scale creatures. We have no idea how the microscopes themselves could have created such detailed, lifelike images. Also, microscopic images offer a continuum of magnifications that connect at the low end with things that we can see with the unaided eye. Increasing magnifications show just what we would expect to see from closer-up views, with features indistinctly seen from further away appearing clearer and more detailed from closer up. Further, the physical theories we use in the design of microscopes, like the laws of optics, are very well understood. Also, microscopes not only permit us to look at the microworld, but to move around in it and manipulate the things we seem to find there. The things we build at the nanoscale work—they have reliable effects in our big-scale world—so we seem to know what we're doing down there. Finally, there are now many different kinds of microscopes and other devices for detecting micro and nano reality. These instruments operate on the basis of very different physical principles, but the images they give us are often in agreement. Van Fraassen's attempt to explain away this fact is a failure.

Does the cumulative weight of these arguments suffice to *prove* that van Fraassen's skepticism is profoundly wrongheaded and that we most definitely do observe micro and nano reality as it is? No, skepticism can never be entirely banished, not even skepticism about ordinary, unaided vision. Philosophers can never eradicate skepticism; they can only hold it at arm's length. However, the upshot of these arguments seems to be that if there is genuine observational knowledge of the physical world, it is arbitrary to say that such knowledge extends only as far as our unaided senses can go.

BUT WHAT ABOUT THINGS THAT ARE *REALLY* UNOBSERVABLE?

Another reason for thinking that van Fraassen is wrong to give such priority to the unaided senses is that the distinction between the observable and the unobservable is not clear. Science writer David Bodanis is sensitive to the thought that something like the salmonella bacterium, that is supposed to exist beyond the range of direct observation, might not be real. He considers and rejects such skepticism:

> Because these creatures are so small it is tempting to think they are not really there, merely some sort of scientific construction. This is false. If you have good eyes you should be able to make out a good range of dust flecks caught in a beam of light in a darkened room. These can be as small as 20 microns (20/1000 millimeter) long. Salmonella are about a tenth of that, so if you had only slightly better vision you could imagine seeing those hairy wriggling submarine life forms all about you. They exist not in a distant, unreal realm, but just a little beyond the normally visible. (Bodanis, 1986, p. 85)

As Bodanis notes, it is easy to imagine a slight enhancement of our vision, perhaps a genetic mutation would give some people the power to see bacteria. For a person with such enhanced vision, seeing bacteria would soon become as ordinary as seeing cats. The difference between the observable and the unobservable therefore seems to be an accident of evolution, and it is hard to see how a strict epistemological distinction can be based on that.

Van Fraassen might be willing to admit that the boundaries between the observable and the unobservable are fuzzy and not fixed. However, he could insist that many scientific theories do clearly postulate entities that are not and will never be observable. For instance, physicists have postulated entities like quarks and superstrings that cannot and probably never will be observed in even the loosest sense of the term. Astrophysicists talk about mysterious dark matter, which, they say, constitutes the bulk of the universe's mass. This dark matter supposedly exerts gravitational influence, but does not otherwise interact with "ordinary" matter. Astrophysicists also postulate the existence of black holes, putative objects that *by definition* cannot be directly observed. Surely, van Fraassen will contend, there is need for skepticism about such hypothetical entities. After all, the only evidence we could have for the existence of such indubitably unobservable things is that they constitute the *best explanation* for the things we can observe. Van Fraassen argues that inference to the best explanation is not a reliable means of establishing the existence of things we cannot observe.

Inference to the best explanation (IBE) is a form of reasoning that we use all the time. Perhaps the most famous instance of IBE in literature is in Daniel Defoe's *Robinson Crusoe*. In the story Crusoe has been shipwrecked for many years on an island that he thinks is utterly deserted. One day while walking on the beach he sees the clear outline of a human footprint. Crusoe does not *see* anybody; he only sees the footprint. Crusoe immediately knows that he is no longer alone on the island. This realization comes so quickly that it is not a process of conscious reasoning, but it is an inference nonetheless. If we were to spell it out, the reasoning would probably go like this: Here is a very distinctive shape in the sand. The only thing that could explain the existence of such a shape is a human foot. Other hypotheses, like, for instance, that the shape was a random effect of wind and surf, are just too implausible to consider. The only reasonable hypothesis is that a human foot made this shape, and where there is a functioning human foot, there is a human being. Again, reasoning of this sort is used every day, and in those mundane contexts van Fraassen has no objection to it. For instance, he is happy to admit that if we hear certain suspicious patterings and scratchings behind the wainscoting, we could rightly conclude that we have mice. After all, a mouse is a readily observable creature and we can learn firsthand the sorts of noises an active animal of that size is likely to make behind the walls. Likewise, van Fraassen would have no quarrel with those paleontologists who look at dinosaur footprints and infer things about dinosaur behavior. We *Homo sapiens* came on the scene about 65 million years too late to see a dinosaur, but *had* we been around in the Mesozoic, dinosaurs would have been most readily observable. So even a dinosaur counts as an observable in van Fraassen's sense.

The case is entirely different, says van Fraassen, when we try to use IBE to justify belief in unobservables. The problem, he says, is that for IBE to do the job here, to give us adequate grounds for thinking that a theory postulating unobservables is likely true, we would have to have a sort of privileged access. That is, we would have to have some *a priori* way of knowing that the correct explanation of whatever we are trying to explain is probably among the hypotheses we have so far considered. Van Fraassen puts it this way:

> Inference to the Best Explanation is not what it pretends to be, if it pretends to fulfill the ideal of induction. As such its purport is to be a rule to form warranted new beliefs on the basis of the evidence, the evidence alone, in a purely objective manner. . . . It cannot be *that* for it is a rule that only selects the best among the historically given hypotheses. We can watch no contest of the theories we have so painfully struggled to formulate, with those no one has proposed. (van Fraassen, 1989, pp. 142–143)

In other words, IBE may be great for choosing among the hypotheses we have already thought up, but how do we know that we are not just selecting the best of a bad lot? That is, what authorizes us to think that one of the hypotheses we already have is the right one? Maybe all of our hypotheses are way off base and the true explanation is one that never has and maybe never will occur to us. This isn't a problem when we use IBE in most everyday situations. We have independent reasons for believing that (in most circumstances) a human foot *and nothing else* is likely to make the distinctive-looking footprint. But things seem different when we postulate an unobservable entity—let's call it "E"—as the cause of some puzzling fact "F". In this case, because E is unobservable, and we have no prior or independent access to E, there seemingly are no independent grounds for saying that E, *and only E,* could explain F. Maybe some other unobservable entity, "E*", could be the correct explanation for F, and it simply never occurred to us that E* might be the cause. Hence, IBE apparently requires privileged access when it is used to argue for the reality of unobservables. We apparently must have some sort of privileged, prior assurance that the true explanation probably lies among the ones we have so far considered. What could give us such assurance? Could we argue that God, or evolution, has designed the human mind so that it is likely to light upon true hypotheses?

Does IBE require privileged access when it is used as evidence for the reality of unobservables? Let's consider the case of black holes. Black holes, if they exist, are among the strangest things in the universe. The best way to begin to think about black holes is to consider what happens to stars as they age and die. Here's the account astrophysicists give: A star's fate is determined by its mass. A star with the mass of our sun will have a rather quiet ending: After about 10 billion years of existence, its hydrogen fuel will start to run out and the star will swell to become a red giant. Eventually the outer layers of the star will be shed into interstellar space, creating a beautiful planetary nebula and leaving behind a compact object known as a white dwarf. A white dwarf is a very odd object. It can pack the mass of the sun into a planet-sized sphere. This means that a white dwarf is extremely dense. A typical average density would be 35,000 grams per cubic centimeter, 35,000 times the density of water and 3,000 times the density of the matter at the earth's core. Since a white dwarf no longer generates its own energy through thermonuclear reactions, it slowly cools over billions of years until it becomes a "black dwarf," a cold, giant crystal harder than diamond.

The fate of massive stars, those significantly larger than our sun, is much more spectacular. They end their lives in cataclysmic supernova explosions.

When a supernova occurs, a star in its death throes can briefly outshine the total luminosity of the hundreds of billions of other stars in its galaxy. If the remnant left behind after a supernova explosion has a mass between about 1.4 and 3 solar masses, that residual star matter will become a neutron star. A neutron star is a very bizarre object with the inconceivable density of a hundred million metric tons per cubic centimeter. Neutron stars spin very rapidly, as fast as hundreds of times per second, thus creating magnetic fields a trillion times stronger than the earth's. These ultra-strong magnetic fields generate beams of microwaves that rotate with the star. When such a beam sweeps past an earthly observer it is detected as a microwave pulse, so rapidly spinning neutron stars are called "pulsars."

The most extreme fate befalls those stars so massive that the supernova remnant has a mass more than three times that of the sun. With such great masses, the gravitational collapse is complete, and, according to the theory of general relativity, everything collapses into a dimensionless point called a "singularity." Within a singularity, according to general relativity, the density becomes infinite and space-time itself is infinitely distorted. Under such conditions the very laws of physics break down. The singularity is shielded from the rest of the universe by an "event horizon" (that nature always decently forbids "naked" singularities by hiding them behind an event horizon is called the "cosmic censorship hypothesis"). The event horizon marks an absolute boundary. Once anything—*anything*—passes inside the event horizon, it is trapped forever. No force in the universe can extricate it. Not even light can escape from inside the event horizon since the gravitational field is so intense that not even the speed of light, the fastest possible speed, achieves escape velocity. Essentially, the event horizon marks a hole in space, and since nothing, not even light, can escape the hole, it is absolutely black.

Because black holes cannot emit or reflect light or any other kind of signal, they cannot be directly observed. Even indirectly detecting a solitary black hole, perhaps as a spot of utter emptiness, would be very unlikely because they are quite small and very far away. Why then are many astrophysicists, who are certainly hardheaded empirical types, nevertheless convinced that black holes really exist? General relativity predicts that such things could exist, but why do so many scientists think that they really do exist? They appeal to IBE, that is, black holes seem to be the best explanation for some things that we do observe. Do astrophysicists, as van Fraassen charges, have to claim privileged access? Do they illicitly assume that the truth is probably to be found in the hypotheses they have so far considered? Let's look in some detail at their reasons for believing that there really are black holes (for the sake of brevity, I'll only consider the evidence for stellar-mass black holes, and not examine the very intriguing evidence for supermassive black holes in the centers of galaxies).

Cygnus the Swan is a magnificent constellation that is high overhead on midsummer nights in the northern hemisphere. At a point within the constellation, astronomers have identified an intense x-ray source they unimaginatively call Cygnus X-1. This source is about 8,000 light-years away and emits x-rays as powerful as the total luminosity of 10,000 suns. Telescopic observation reveals that Cygnus X-1 is a supergiant star in a binary system with an unseen companion. Astrophysicists have inferred that the best explanation of such intense x-ray emission is the existence of a black hole as the unseen companion of the star. The gravitational acceleration into the black hole of material stripped from the companion star could create the energies needed. The intense gravitational field would pull material from the companion star that would accumulate in a rotating accretion disk around the black hole. Compression and friction within the disk would raise the temperature of the accreted material to a million degrees centigrade, which would generate the observed x-rays.

The problem is that the energy necessary to create the x-rays could also be created if the unseen companion were a white dwarf or a neutron star. However, variations in the x-ray emissions on the order of a millisecond indicate that the unseen component of Cygnus X-1 must be less than 300 km in diameter. Light travels only 300 km in a millisecond, so any variation in x-ray emissions that takes only a millisecond to occur cannot come from a body larger than 300 km. Since a white dwarf is about earth-sized, this shows that the unseen component cannot be a white dwarf. Also, a law discovered by Johannes Kepler in the early seventeenth century allows astronomers to place lower limits on the masses of bodies in orbit about a common center of gravity. These calculations indicate that the compact object in the binary system has a mass of at least 3.4 solar masses, well above the upper limit of 3.0 solar masses for a neutron star. Most astrophysicists appear to think that the invisible companion in the X-1 binary system is probably a black hole. Some disagree, but they dissent because they doubt some of the calculations of distances and masses, not because they reject the underlying reasoning.

Now van Fraassen would probably agree that of the available hypotheses—white dwarf, neutron star, or black hole—the black hole hypothesis is the best of the lot. But he would deny that astronomers have any grounds for further asserting that the black hole hypothesis is *probably* true. After all, how do we know that the invisible component of Cygnus X-1 is not some sort of object never imagined by theorists?

Van Fraassen is certainly right that the invisible object in Cygnus X-1 could be something we never imagined. Maybe the whole edifice of theory and observation that leads astrophysicists to believe in black holes, neutron stars, and white dwarfs is wrong. However, to say this is to say nothing more than that scientific knowledge is *always* tentative and fallible—something everybody admits. Unless we have some reason to think that our most careful observations and the predictions of our best-confirmed theories are not trustworthy, and not just that they might not be, there cannot be anything objectionable about basing our probability judgments on what those theories predict and what those observations seem to show. Observation shows that one component of the Cygnus X-1 binary system is invisible so it cannot be an ordinary star. Further observations and measurements appear to show that the object is very small and very massive. General relativity, one of the best-confirmed physical theories, predicts straightforwardly and in considerable detail what will happen to matter when it achieves such densities. What it predicts is that a singularity will form surrounded by an event horizon of predictable radius. In other words, a black hole will form.

Put simply, astrophysicists think that the invisible component of Cygnus X-1 is a black hole because, given everything they think they know about the universe, it just couldn't be anything else! It can't be a white dwarf or a neutron star, so a black hole is all that's left. Do astrophysicists implicitly appeal to special privilege to limit the candidate hypotheses to just these three—white dwarf, neutron star, or black hole? No, these three, and only these three, are the candidates permitted by the total state of our presumed astrophysical knowledge at this time. Astrophysicists think they have good reasons for holding that all stars below a certain mass will eventually form white dwarfs or neutron stars and not anything else. Further, according to well-confirmed theory, anything denser than a neutron star will collapse into a singularity to form a black hole. There is no implicit appeal to privileged access.

We can summarize the reasons an astrophysicist might give for believing that black holes probably exist: We observe cases like Cygnus X-1 (other binary systems that are *better* black hole candidates have been more recently discovered) where we can see an ordinary star orbiting a common center of mass with an unseen companion. Further observation and inference (where the inferences are based on indisputably correct premises, like the value of the speed of light or the reliability of Kepler's third law) show that the unseen companion is too large to be a white dwarf and has a mass beyond the theoretical limit for neutron stars. We think that when matter achieves certain densities, a singularity will form. We think this because it is a straightforward prediction of the theory of general relativity, a core theory for astrophysicists. Therefore, of the three possible invisible causes of the massive x-ray emissions detected in Cygnus X-1, the best explanation is a black hole.

We see that IBE enters in only after a great deal of theory and observation have given us reason to delimit our candidate hypotheses to three and only three—white dwarf, neutron star, or black hole. The decision to limit our candidates to these three is therefore apparently based on *scientific* grounds and does not appeal to any sort of privileged or *a priori* knowledge. It therefore seems that van Fraassen is wrong to think that IBE must always appeal to privileged access when it concludes that an unobservable entity probably exists. There are often good *scientific* reasons for thinking that the right explanation lies within the batch of hypotheses under consideration. Maybe van Fraassen was right that we risk less if we only expect our theories to be empirically adequate and refrain from asserting their truth. But when it looks like we *can* get some truth, why not go for it?

Having taken a rather lengthy detour through astrophysical details, let's pause to remind ourselves of van Fraassen's initial objection to IBE: IBE can only select from among the hypotheses we happen to have thought of, and there is no way, short of claiming privileged access, that we can claim to know that the true explanation probably lies within the batch of hypotheses we actually have. Privileged access, once again, would mean that we have some sort of prior assurance that true explanations are likely to be among the ones we just happen to think up. The only way to argue for such privileged access would be to claim that our minds are somehow designed (by God or evolution) so that the hypotheses that we think up are likely to contain true explanations. Having placed this heavy burden of proof on realists, defenders of constructive empiricism can enjoy the smug confidence that realists are unlikely ever to establish such recondite theses. But if, as we have seen, it is the straightforward work of science, and not privileged access, that assures scientists that the true explanation probably lies within a small set of candidates, then the burden of proof shifts to the critics of IBE. Such critics would now have to show why the use of IBE in astrophysics is really in principle different from its use when we say that there is probably is a mouse behind the wainscoting. Whether it is mice or black holes that best account for our data, we seem to have good reason to think that we have gotten the *right* explanation.

Of course, we have just scratched the surface of the realism/antirealism debate here, and antirealist philosophers would stoutly contest the conclusion I have just drawn. There are many other arguments and issues relevant to the debate that we do not have room to examine. At the present time among professional philosophers of science this debate is at a stalemate. Neither side seems to be capable of gaining a decisive advantage over the other. This does not mean that the realism/antirealism debate is going to go away. It has been around for thousands of years and it is an issue that spontaneously arises from the nature of scientific inquiry. Indeed, any attempt by scientist, philosopher, or theologian to penetrate to the reality behind the appearances is bound to invite a skeptical riposte.

SO, WHAT REALLY IS THE GOAL OF SCIENCE?

If we do reject the claim that the goal of science is merely to save the appearances, should we say that science aims to generate theories that tell us the truth about the natural world—the view van Fraassen attributes to scientific realists? Well, it is no doubt misleading to speak of "the" goal of science, since science has many goals, but it is intuitively plausible to say that science seeks truth. Surely scientists would not insist that theories be tested so rigorously if they did not care for truth.

Truth is important in science, but the pursuit of truth is subordinate to the goal of *understanding*. Sometimes we understand something by learning the literal, exact truth about it—but not always. Sometimes we cannot know what something *is;* the best we can do is to say what it is *like*. Often in understanding some aspect of the natural world we have to use models, analogies, and metaphors rather than literal descriptions. Sometimes scientists make physical scale models—like the plastic and metal model that Francis Crick and James Watson built to guide them in uncovering the double-helix structure of DNA—but these are not the kinds of models meant here. Here the kind of model we are talking about is an abstract, usually mathematical representation that might not give any sort of *picture* of what it represents, but only symbolic tools for talking about that hidden reality. Other models will offer only a very simplified or idealized depiction of some part of the natural world. Such a model is not a portrait of reality, but is more like a sketch, or even a caricature that emphasizes only the most scientifically crucial aspects while omitting all unnecessary detail—including all those crucial details that often make those objects so fascinating for us. As far as gravitational theory goes, it does not matter at all whether a mass in free fall is a lead sphere, a brick of gold bullion, or an NBA superstar. So, when our theories employ such abstract and simplified models, the theory does not aim to give a literally true description of the world.

In fact, science could hardly work if we always insisted that theories tell the exact truth. Consider the law of free fall discovered by Galileo. Mathematically, the formula is $d = gt^2/2$, that is, the distance (d) traveled by a freely falling body, over a time (t), is equal to the acceleration due to gravity (g), times the square of the elapsed time (t), with the product divided by two. Since the acceleration due to gravity is 32 feet per second squared, if a skydiver jumps from a plane and is in free fall for four seconds, he or she will fall 256 feet, according to this law. The discovery of this law was a great landmark in the history of science. It is still found in all elementary physics textbooks. The problem is that it is not true. Strictly speaking, the law holds only for objects falling in a total vacuum, and total vacuums do not exist naturally (not even in interstellar space) and cannot be created in laboratories. In real-life situations, Galileo's law is only an approximation (often a very good one). In fact, if you jump from a plane you probably will not fall *exactly* 256 feet in four seconds. If we insisted on an absolutely true law for real-life falling bodies, we would have to take into account the very complex influence of factors like the effects of air resistance on bodies of various shapes, and the influence of ambient temperature and humidity. Such "laws" would be far too complex to use.

The upshot is that the laws of nature as scientists formulate them are a compromise between truth and usefulness. The complexity of exact truth has to be sacrificed for the virtues of simplicity and universality. This is why the laws of physics so often postulate ideal entities like total vacuums, frictionless surfaces, and point-masses. It is not that the laws of physics lie; they do not even fib. Rather, they constitute what philosopher Stephen Toulmin calls "ideals of natural order"—representations of what nature would be like under ideal circumstances (recognizing, of course, that circumstances are never actually ideal). Such ideals function as baselines for explaining the behavior of real entities. To the extent that things behave like the ideal, that behavior is considered natural and in need of no further explanation. Behavior that departs *too* far from the ideal requires further explanation (*slight* departures from the ideal are expected). This is not just true of the laws of physics, by the way, but also those of biology. Mendel's genetic theory predicts ratios of the various phenotypes that are ideally expected in a given generation, but the ratios actually found are very unlikely to match precisely with the ideal value. In fact, if the actual experimental value matched the predicted value *too* closely, we would suspect that the experimenter had fudged the data!

Realists often defend their views by asserting that scientists spontaneously take a realist view of theoretical entities. This is a plausible claim. Generally, physicists seem to think that there really are quarks; chemists think there really are atoms; biologists think that DNA really does have a double-helix structure, etc. However, it does not follow that scientists automatically hold that getting literal truth is *the* goal of science. Scientists take a more pragmatic view and are often quite willing to say, as did

Newton, "*hypotheses non fingo!*" when a theory works but nobody knows why it does. A case in point is quantum mechanics. There is no more successful theory in the history of science than quantum mechanics. Hundreds if not thousands of tests have never turned up any evidence against it, and its use is so basic to contemporary physics that the field would not exist without it. Yet quantum mechanics, for all its usefulness, presents an impenetrable mystery.

As represented in the mathematical language of quantum mechanics, there are certain properties of quanta—the tiniest bits of matter—that are given no definite values until they are actually measured. The so-called "dynamic" properties of quanta—such as position and momentum—are represented as being in no definite state prior to measurement. The values of dynamic properties for unmeasured quanta are represented as a "superposition of states," a mixture of possible states rather than a definite value. Only after a detecting device actually performs a measurement does the property take on a definite value. Now nobody knows what is going on here. The standard interpretation, called the *Copenhagen interpretation,* says that it is not just that we *don't know* the values of those dynamic properties prior to measurement; there *are* no values before we test for them! When we interact with those quanta in order to measure them, then—and only then—do they take on definite values. Nobody understands how the act of measuring quanta could cause indefinite properties to become definite. This is the famous "measurement problem" of quantum mechanics.

Does it bother most physicists that nobody knows what is really going on here? Not at all. Of course, many other interpretations have been offered that try to make sense out of all this, but most physicists seem to think that such interpretations are more philosophy than science since they have nothing to do with the use or application of quantum theory. The observed quantum facts are the same on all such interpretations. As one wag put it, the average quantum mechanic has no more interest in philosophy than the average auto mechanic. The working scientist accepts quantum mechanics because it *works;* it tells us what we want to know and lets us do what we want to do. The fact that it doesn't tell us what is *really* going on when we measure a particle's dynamic properties is not held against it any more than Newton's failure to explain why gravitational force exists kept people from using his law of universal gravitation. Maybe we will never know the whole story about what is really going on in the quantum world, and we will have to be satisfied with a theory that leaves such gaps.

Clearly, therefore, the statement that truth is *the* goal of science is simplistic and misleading. We often have to use models, metaphors, and analogies to understand things that we cannot describe literally; sometimes these give us our *only* tools for grasping things. Other accepted theories represent idealizations or simplifications that are not literally true of the actual world. In particular, the laws of nature as we formulate them strictly apply only to idealized situations that are never perfectly realized. Further, theories like quantum mechanics often fail to tell us some of the things we would really like to know, but they do provide invaluable mathematical tools for explanation, prediction, and control. A good scientific theory enhances our understanding of the world, and understanding sometimes is and sometimes is not the same thing as knowing the literal, exact truth about things.

In fact, all of the talk about metaphors, analogies, and models may have left you wondering whether there is much difference between scientific realism and constructive empiricism if the best we can often do is to present an abstract model of unobservable entities. Hasn't the realist really largely given up the game and sided with the antirealist? The realist has to admit that we cannot have a literal description of the dynamic properties of unobserved electrons. The best we can presently do, and maybe the best we will ever do, is to make an abstract mathematical model (the "wave function") of the electron that we use to make reliable predictions about what we will probably find when we do perform a measurement. As we note above, what *really* is going on with unobserved electrons is anybody's guess. (The realist can adopt a realistic *interpretation* of what is going on at the quantum level, but, again, this is an interpretation added to the theory, not an element of the theory itself.)

So, the scientific realist has to admit that even our best theories in even our most "mature" sciences often represent aspects of nature in ways that are not even truth-like. On the other hand, constructive empiricists do not deny that there *is* an unobservable reality, and they have to admit that in case after case the physical world acts exactly *as if* that unobservable reality were constituted of unobservable particles of specific types. In other words, constructive empiricists admit that the best models of the unobservable structure of the cosmos are precisely the ones that postulate electrons, protons, quarks, etc. So, at least in dealing with the postulated particles of fundamental physics, both camps seem to be saying something tantamount to this: "There exists a level of reality far too small to be directly observed. We cannot say exactly what that reality *is*, but we can construct abstract mathematical models of that reality that reliably predict what we will experimentally observe."

It would be nice to accept the irenic suggestion that at rock bottom not much separates the scientific realist from the constructive empiricist, but the proposed reconciliation won't wash. Though scientific realists might admit that we do not and maybe cannot know everything about electrons, they do hold that "electron" names a natural kind, that is, that our concept "electron" corresponds to a real category of thing in the natural world. In other words, the realist thinks that when we classify some things as electrons, we are classifying them as they *are*. Further, they hold that there are very many particular instances of the electron kind. As philosophers like to say, there exist many "tokens" (individual electrons) of that "type" (the kind of particle we name by the term "electron"). They also hold that there are some properties of electrons, the so-called static properties such as charge and mass, that we can say that electrons always have under all circumstances, whether they are being observed or not. Constructive empiricists do not think we can make any such positive assertions about what sorts of natural kinds exist in the unobservable realm. *All* we can say is that things can be modeled in certain ways and that those models make reliable predictions about observable consequences, so there is an irreducible difference between scientific realism and constructive empiricism.

Taking into consideration the entire discussion of rationality, realism, and antirealism through this whole book, I offer the following definition of scientific realism:

> Scientific realism is the doctrine that makes the following three claims: (1) There is a physical world, with determinate properties, that exists independently of our perceptions or understanding of it. (2) Science aims to give us, in *some* of its theories, a literally true picture of aspects of the physical world. This aim extends to giving a literally true picture of some of the unobservable parts of the world. It is eminently reasonable to regard well-confirmed theories that have that aim as truth-like. (3) Further, science has *actually succeeded* in identifying and accurately describing some of the natural kinds of observable and unobservable things in the world.

Scientific realism first makes a *metaphysical* claim, that there exists a real physical world "out there" that is the way it is independently of how we perceive it or think about it. This may seem patently obvious to many and not worth stating. Recall, however, that some of the people discussed in earlier chapters did seem to question this basic assertion. For instance, in places Kuhn seemed to be saying that the world that exists changes from paradigm to paradigm. Also, Bruno Latour and Steve Woolgar in *Laboratory Life* and elsewhere really do sometimes seem to question the existence of a determinate, independent physical reality. Scientific realists affirm that there is such an objective physical reality.

Scientific realism also makes an *epistemological* claim, namely, that it is entirely reasonable to regard some well-confirmed theories as truth-like.

Scientific realists affirm that when a theory aims to tell us the truth about some aspects (observable or not) of the world, and when that theory is supported by sufficient evidence or passes sufficiently stringent tests, we are warranted in regarding those theories as approximately true. Concomitantly, it is a legitimate goal of science to try to discover as much truth as possible about the universe, including truths about its deep, unobservable structure. Modestly asserting that seeking theories that are literally true is only *one* goal of science is consistent with the possibility that theories that are merely empirically adequate may sometimes give us our best possible understanding of some natural phenomena. A modest scientific realism therefore rejects the stark choice offered at the beginning of the chapter—whether we want our theories to be true or merely to have true consequences. We now see that "truth or consequences" is a false dilemma. In a nutshell, the scientific realist thinks that science should strive to maximize our understanding of the world, and holds that this project will often involve the discovery of truths about unobservable aspects of the natural world. For the scientific realist, the problem with all forms of antirealism is that, by refusing to believe what science tells us about the unobservable structure of the world, these doctrines settle for *less* understanding than we can get.

Scientific realism as I formulate it also claims that science really has succeeded in correctly identifying and describing some of the constituents of the natural world, even some that are unobservable. Some scientific realists shy away from tying their position to any particular body of scientific claims. If we do, and those claims are later shown wrong, what happens to scientific realism? However, it seems to me that if we do not insist that science really has succeeded to some extent in identifying and accurately describing (i.e., describing with a high degree of truth-likeness) some of the actual constituents of the world, we concede the whole case to the pessimistic meta-induction. If science hasn't gotten anything right so far, then the main epistemological claim of scientific realism—that we are warranted in regarding some theories as truth-like—would be in doubt. Finally, to distinguish scientific realism from the "blind realism" mentioned in the last chapter—where we affirm that many scientific theories must be right but we are noncommittal about which ones actually are—many scientific realists may wish to add to the above definition that we *do* have good grounds for identifying some *particular* theories as truth-like.

Having formulated and defended my definition of scientific realism, it follows only to recommend it to the reader as the best philosophical perspective on science. As I see it, modest scientific realism of the sort defined above is an attractive *via media,* a middle course, between philosophical extremes. It does not make the hyper-realistic claim that truth is *the* goal of science; it concedes that our best theories sometimes might be merely empirically adequate or employ models, metaphors, or analogies that cannot give us a literal picture of things. On the other hand, modest scientific realism thinks that antirealists go too far in recommending that we *never* believe theories about the unobservable parts of the universe. Such a view leaves us with an impoverished understanding of the world. Finally, by affirming the actuality of an independent physical reality, scientific realism resists the claims of social constructivists, postmodernists, and relativists that, in various ways, make physical reality an artifact of our concepts. Even if it is coherent to claim that reality is an artifact, and I'm not sure that it is, it is an extremely counterintuitive notion for many people. I think that John Searle is right that the "default setting" of the human mind is to think that there is a world "out there" that exists independently of us. Those who are not persuaded by the arguments for changing that default setting will favorably regard scientific realism's affirmation of the independence of physical reality.

When you stop and think about it, it really was incredibly brash for people to start thinking that they could come to understand the hidden inner workings and structure of the amazing and beautiful, but extremely complex and mysterious cosmos that we are a small part of. It is a brashness that can only be captured by that wonderful Yiddish word *chutzpah.* Maybe the biggest gamblers in history were the pre-Socratic Greek philosophers who had the wild idea that the human mind could penetrate to the inner essence, the *arche,* of things—the reality underlying our world of shifting appearances. Of course, each philosopher gave a different answer about what that underlying reality was supposed to be, but they agreed that there was such a reality and that the human mind could discover it.

Now, 2,500 years later, the pre-Socratics' big gamble seems to have paid off—spectacularly. Natural science now offers a rich set of powerful theories that purport to tell us about unobservable parts of the world. These theories enjoy the support of meticulously gathered evidence and have survived the most rigorous attempts to refute them. But have they earned the right to be *believed,* or does it still require *chutzpah* to believe that there are (for instance) electrons and that we know some literal truths about them? In these last two chapters I have given some reasons why I think that we are not going too far out on a limb to think that science does sometimes tell us the real story about the unobservable things in the universe. But issues such as this have been debated *ad nauseam* by philosophers of science, and, as I said earlier, the realism/antirealism debate is deadlocked.

Maybe some progress could be made if we cast our net somewhat wider, that is, if we placed the realism/antirealism debate in the context of the broader epistemological question about how, in general, we could have warranted beliefs about unobservable things. At the beginning of this chapter I noted that humans often believe in many things, besides the theoretical entities postulated by science, that they cannot detect with their senses, and that these claims raise their own questions about what we can rationally believe when we move beyond what we can observe. For instance, I asked what it would take to make it reasonable to believe in God even if He doesn't make spectacular public appearances. Perhaps then we could make progress in the philosophy of science *and* the philosophy of religion if we could get clearer on some basic epistemological points, like how, in general, claims about unobservables (whether gods, ghosts, or gravitons) might be warranted. It seems to me that the most promising line of inquiry here would be to examine more deeply the nature of IBE, inference to the best explanation. Philosopher Peter Lipton (1991) and others have written engaging and enlightening works on this subject, and I would like to pursue these issues in detail. However, to do this the right way would require me to write another book.

FURTHER READINGS FOR CHAPTER FIVE

Norwood Russell Hanson was a brilliant young philosopher of science who was killed in a flying accident at the age of forty-three. His tremendously amusing essay "What I Don't Believe" is found in *What I Do Not Believe, and Other Essays,* edited by Stephen Toulmin and Harry Wolf (Dordrecht, Holland: D. Reidel, 1971), pp. 309–331.

The story of the discovery of the electron is related clearly and authoritatively in *The Discovery of Subatomic Particles,* revised edition, by Nobel Prize–winning physicist Steven Weinberg (Cambridge: Cambridge University Press, 2003). Weinberg is a top-notch scientist who also is an outstanding writer of science for the general public. He tells the story of Walter Kaufmann, the German physicist who did the same sort of experiment Thomson did, only better, in his *Dreams of a Final Theory* (London: Vintage, 1993).

Jean Perrin's 1913 book *Les atomes* was published as *Atoms,* translated by D. Hammick (Woodbridge, Conn.: Ox Bow Press, 1991). Perrin's work was the subject of a distinguished study by Mary Jo Nye titled *Molecular Reality: A Perspective on the Scientific Work of Jean Perrin* (New York: American Elsevier, 1972).

Bas van Fraassen's best-known work is *The Scientific Image* (Oxford: Clarendon Press, 1980). Van Fraassen stated his constructive empiricism in this book and thereby kicked off what was probably the liveliest debate in the philosophy of science since the publication of Kuhn's SSR. By 1980 philosophy of science had finished a long struggle to throw off the dominance of logical positivism, a view that had been strongly influenced by Mach's instrumentalism. Then van Fraassen came along offering a new form of antirealism, constructive empiricism, supported by a very challenging set of arguments. Van Fraassen offered additional arguments against realism in *Laws and Symmetry* (Oxford: Clarendon Press, 1989). Van Fraassen is a clear writer, but he often gets into quite technical issues. An introduction to his views for beginners, written from a sympathetic perspective, is *Understanding Philosophy of Science* by James Ladyman (London: Routledge, 2002). The arguments of van Fraassen's critics, with a reply by van Fraassen, were brought together in *Images of Science: Essays on Realism and Empiricism,* edited by Paul M. Churchland and Clifford A. Hooker (Chicago: University of Chicago Press, 1985). A pointed critique of van Fraassen's and other forms of antirealism is Michael Devitt's *Realism and Truth,* second edition (Oxford: Blackwell Publishers, 1991).

Niels Bohr's antirealism is examined in Helge Kragh's *Quantum Generations: A History of Physics in the Twentieth Century* (Princeton: Princeton University Press, 1999). This is an excellent, nontechnical history of physics in the twentieth century.

Francis Crick tells the story of neuroscience in elucidating the process of seeing in his *The Astonishing Hypothesis: The Scientific Search for the Soul* (New York: Simon & Schuster, 1994). Crick's enthusiasm is unbounded and infectious. Seeing comes so automatically for us that it comes as a shock to realize just how astonishingly complex the process is and how much we still have to learn. Even more shocking are the ways that "seeing" can go so wrong. When people hear that large numbers of sane, intelligent, and honest people have "seen" miracles, UFOs, Bigfoot, ghosts, etc., it is almost irresistible to think that there must be something to all these things. However, Terence Hines in *Pseudoscience and the Paranormal: A Critical Examination of the Evidence* (Amherst, N.Y.: Prometheus Books, 1988), shows how easily we can "see" what isn't there.

A fascinating, and disturbing, look at the microworld in your own house is found in David Bodanis's *The Secret House* (New York: Simon & Schuster, 1986). The pictures of the dust mites and other tiny creatures that live with (and on and in) us are inevitably somewhat disquieting. Bodanis also has the picture of the pin with the bacteria clustered on the tip. It makes you thankful to have a good immune system.

Ian Hacking's essay "Do We See Through a Microscope?" is published from pages 132 to 152 in the book *Images of Science*, mentioned above. In my view, Hacking is one of the most interesting of current philosophers of science, and his arguments for realism are among the most effective.

Another essay by Hacking defending realism is "Experimentation and Scientific Realism," in *Scientific Realism,* edited by Jarret Leplin (Berkeley: 1984), pp. 154–172. If Hacking were writing these essays today, he would probably make use of the recent developments in nanotechnology. A very readable introduction to these amazing developments is *Nanotechnology: A Gentle Introduction to the Next Big Idea* by Mark Ratner and Daniel Ratner (Upper Saddle River, N.J.: Prentice Hall, 2003). Van Fraassen's reply to Hacking is in *Images of Science.*

The inimitable Isaac Asimov wrote about black holes in his usual stimulating and crystal-clear manner in *The Collapsing Universe* (New York: Walker and Company, 1977). A good book by a black hole specialist is *Black Holes* by Jean-Pierre Lumient, translated by Allison Bullough and Andrew King (Cambridge: Cambridge University Press, 1992). If black holes don't give you the creeps, not much in the natural world will. They are fascinating objects; you just have to be glad that, so far as we know, they are very far away.

Stephen Toulmin's *Foresight and Understanding: An Enquiry into the Aims of Science* (New York: Harper Torchbooks, 1961), is a succinct (115 pages) work that, unlike so much other writing about science from the early 1960s, remains fresh, relevant, and insightful. Toulmin really knows the history of science, and this knowledge adds substance to his philosophical conclusions.

Probably the best book for general audiences on the various interpretations of quantum theory is Nick Herbert's *Quantum Reality: Beyond the New Physics* (Garden City, N.Y.: Anchor, 1985). There are loads of books on quantum mechanics written by authors with an ideological ax to grind who want to use QM to justify mysticism, idealism, or occultism. Quite a few of these know nothing about QM; others understand the physics, but put a highly tendentious spin on it. A good corrective for these is Victor Stenger's *The Unconscious Quantum: Metaphysics in Modern Physics and Cosmology* (Amherst, N.Y.: Prometheus Books, 1995). A very readable and interesting introduction to the philosophy of physics that defends a realistic view is Peter Kosso's *Appearance and Reality: An Introduction to the Philosophy of Physics* (Oxford: Oxford University Press, 1998).

The best current book-length study of IBE is Peter Lipton's *Inference to the Best Explanation* (London: Routledge, 1991). Lipton specifically addresses van Fraassen's and other criticisms of IBE.

INDEX